Springer Aerospace Technology

Wolfgang Kitsche

Operation of a Cryogenic Rocket Engine

An Outline with Down-to-Earth and Up-to-Space Remarks

With 100 Figures and 19 Tables

Springer

Wolfgang Kitsche
Deutsches Zentrum für Luft- und
Raumfahrt e.V. (DLR)
Langer Grund
74239 Hardthausen
Germany
wolfgang.kitsche@dlr.de

ISBN 978-3-642-10564-7 e-ISBN 978-3-642-10565-4
DOI 10.1007/978-3-642-10565-4
Springer Heidelberg Dordrecht London New York

Springer Series in Aerospace Technology ISSN 1869-1730 e-ISSN 1869-1749

Library of Congress Control Number: 2010932446

Cover design: WMXDesign GmbH, Heidelberg

Printed on acid-free paper

Springer is part of Springer Science+Business Media (www.springer.com)

... and one remark up to heaven

All we do should increase the honour of God, as a beneficial lead for those who have lost their connection to their Creator and Father in heaven and to the joy of those who know their Lord and Saviour; only then is a service complete. Because we cannot complete any service without His help, prayer is the beginning of any deed and service. All human measures, planning and skill are no guarantee for the success of anything, but the help of God can lead from situations without hope to wonderful success; for Him nothing is impossible.

Wolfgang Kitsche, June 2009

... by my God have I leaped over a wall.
King David, Psalm 18 Verse 29

Preface

"A Test Facility, if you've seen one of them, you've seen all of them"; that was the comment of a colleague showing a picture in his presentation about the utilisation of hydrogen. This book will invite the reader to see not only the trivial pipeline on the test facility, but rather to discover that here all efforts are merging to operate a propulsion system of ultimate performance. The development and studies of many branches of science, constructions and prototypes of various high tech companies have to prove their quality here. A prototype of some gigawatts is handed over to the test engineer with a value of umpteen millions of Euros and the focus of the whole space program is on the result of his work. The test cell of the facility is the meeting point of specialists in the most exciting phase of the development of a rocket engine.

Some authors having devoted a chapter of their books to the test facility, this book sets the focus on the operational aspects of the rocket engine in the test facility. This book will be useful to engineers and scientists who are concerned with the test facility, to aerospace students it will provide an insight to the job on the test facility and to interested readers it will provide an impression of this thrilling area of aerospace.

The comments in this book reflect the experience of 2 decades of test periods on one of the two largest test facilities for cryogenic rocket engines in Europe.

Hardthausen, Germany
October 2010

Wolfgang Kitsche

Acknowledgement

Thanks go first of all to everybody in the test team – only their elaborate work made so many successful tests possible. The author also wants to thank his colleagues from the engine manufacturer SNECMA for the amicable cooperation which amplified the ability and professionalism of the whole test team. Special thanks go to Oskar Haidn and Gerd Krühsel for their friendly support; they closely studied the draft of this book and addressed many hours to discussing and improving the content. Thanks are also due to the head of the DLR Space Propulsion Institute, Stefan Schlechtriem and the former Head of Institute, Wolfgang Koschel for their support in the course of producing this volume. Many thanks also go to the author's family who had to abstain from his presence for many hours but always had the higher priority.

The greatest honour is due to God, who gave us all the ability and the opportunity to do this thrilling work without any serious incident.

Hardthausen, Germany Wolfgang Kitsche

Contents

List of Abbreviations

A/D	Analog/Digital
ARTA	*Ariane* Research and Technology Accompaniment
CNES	Centre National d'Études Spatiales
CTA	Commission de Traitement des Anomalies
DLR	Deutsches Zentrum für Luft- und Raumfahrt
EADS	European Aeronautic Defence and Space Company
ECA	Evolution Cryotechnique Type A
EGAS	European Guaranteed Access to Space
ESA	European Space Agency
FMECA	Failure Mode Effects and Criticality Analysis
FORTRAN	FORmula TRANslation, program language
H0	Time of hot run start
H2	Time of hot run stop
He	Helium
L5	Launch number 5
LH2	Liquid hydrogen
LO2	Liquid oxygen
LOX	Liquid oxygen
LSD	Load Simulation Device
MCC	Measurement, control, command
MkII	Mark II, initial name of the *Vulcain 2*
N_2	Nitrogen
NASA	National Aeronautics and Space Administration
NCR	Non-conformance report
NE	Nozzle extension
NEA	NetzErsatzAggregat
Norsk Data	Norwegian computer system
NPSH	Net pressure suction head
O_2	Oxygen
P1.2	Test facility on the Lampoldshausen test centre
P4.1	Test facility on the Lampoldshausen test centre
P5	*Vulcain* test facility at DLR, Lampoldshausen
P8	Test facility on the Lampoldshausen test centre

PF50	*Vulcain* test facility at SNECMA, Vernon
PLC	Programable logic controller
RAMS	Reliability, availability, maintainability, safety
SCA	Attitude control system
SEP	Société Européenne de Propulsion
SMO	Specification de mise en ouvre
SNECMA	Société Nationale d'Etudes et de Constructions de Moteurs d'Aviation
SPF	Single point failure
TNT	Trinitrotoluol
USV	UnterbrechungsfreieStromVersorgung
VFW	Vereinigte Flugtechnische Werke

Chapter 1
Introduction

A century ago the first aircraft took off and aviation started to make tremendous progress. Half a century ago we succeeded for the first time in sending a man into space. This great step excited hopes of being able to travel to the planets of our solar system within the next few decades. But in astronautics this rapid evolution was not repeated. Today the transport of payload from our precious planet into orbit around it or into space beyond is still a huge challenge for men.

In aviation, air breathing propulsion systems are applicable for flights up to around 30 km and less than 3 km/s. Innovative electrical propulsion systems for in-space flight have proven their efficiency in space and their ability to maintain and control the orbit velocity of the spacecraft. But for the gap between these two propulsion systems, only the rocket engine with a chemical process is available.

The fact that all propellants for the flight have to be onboard necessitates a huge launcher with high take off mass compared to the mass of the payload. Hence a space mission is drastically more expensive than any other transport. High reliability despite high load is therefore a requirement for all components of a launcher system.

The performance and reliability of the propulsion system is of essential importance for any space mission. A failure of the propulsion system mostly leads to destruction of the whole launcher and loss of the complete mission. Fifty percent of lost launcher missions have their origin in a failure of the propulsion system.

Great efforts have been made to improve the performance and reliability of chemical propulsion systems. Special materials were developed, the design of the components of the rocket engines were optimised, the behaviour of the propulsion systems was carefully scrutinised and the theoretical prediction of the performance was improved. However, the prediction of the behaviour of this complex system is still very difficult, since it requires the knowledge of dynamic processes, transients and interaction of various subcomponents of the propulsion system.

Many remaining questions and the confirmation of all calculations are left to the test of the rocket engine on a test facility. Here we can study the real characteristics of the engine, we can subject it to the limits of its performance, and we can observe which traces under the load of repeated long duration tests will occur on the engine subcomponents. The test facility is an indispensable tool to bring a propulsion system to maturity for flight and is therefore a most interesting place of work for space scientists.

W. Kitsche, *Operation of a Cryogenic Rocket Engine*, Springer Aerospace Technology,
DOI 10.1007/978-3-642-10565-4_1, © Springer-Verlag Berlin Heidelberg 2011

Normally a test facility is specially designed for the very rocket engine to be developed and qualified. Each test preparation begins with an intensive look at the details of the propulsion system. Its characteristics and the test requirements have to be matched with the performance of the facility. Test preparation, execution and post-test activities are laid down in a system of procedures performed by a specially trained test team. The procedures not only cover the handling and testing of the rocket engine but also the operation of various bench subsystems. All bench systems have the central aim of enabling the engine test and to simulate an environment to the tested rocket engine as if it were on the real launcher.

As well as the launcher, the test facility is a very valuably module for astronautics. In spite of the high investment, the combined experience and know-how of the test facility and its team is of even higher value. The extraordinary, almost unique process of running a test facility for rocket engines and the gaining of experience from operating the facility and the rocket engine have to be combined and conserved in order to build up the art of operating rocket engine test facilities as a reliable module within space science. This book is an outline of how to operate a test facility, how to maintain it and keep it available and how to modify it over the decades to meet the needs of rocket propulsion technology.

Chapter 2
Operational Aspects of the Rocket Engine and the Test Facility

Design, construction and production are essential phases in the development of a rocket engine. As soon as the engine subcomponents are at a satisfactory development status, a complete engine, ready for operation, is examined on a test facility which is specially designed for this very engine. The aim of an engine test is to check and optimize the operational cycle of the engine and finally to confirm the readiness of the rocket engine for use on the launcher. When the rocket engine is used during flight, or when it is tested on the ground facility, the *operational aspects of the rocket engine* have to be considered. This means the engine dynamic behaviour during *start up*, during transit from one to another operational point, the characteristics at extreme operational points and during *shut down* of the engine.

The bench systems are adjusted to this *operational cycle*. The bench supplies the engine with propellant, secondary fluids (see Appendix J), electrical power and control signals. The operation on the facility also requires simulation of the conditions which will be met later on during actual flight. Furthermore, conditions have to be handled which have their origin in the engine operation itself (e.g. noise, exhaust jet, risks). In a wider sense, the safety of the engine environment is also part of the operational aspects as well as the measurement of the physical parameters. These measurements are necessary for the regulation, for monitoring of important engine parameters, for evaluation of the engine test and for evaluation of the engine behaviour during flight. Running the required bench systems and procedures to resolve these tasks is the subject of the *operational aspects of the test facility*.

An engine test can also be executed on the launch pad of the rocket (Fig. 2.1). In this case the test objective is the validation of the launch preparation related to the engine and its interaction with the stage (see Appendix E, Fig. E.1), including an ignition and a regular shut-down.

2.1 Propulsion Systems

The main components of a propulsion system of a rocket are the rocket engine, the fuel tank, the oxidiser tank and a pressurisation system. The first and second stages of a launcher basically consist of almost only the propulsion system. Even on satellites the propulsion system forms the major part of the weight and structure.

W. Kitsche, *Operation of a Cryogenic Rocket Engine*, Springer Aerospace Technology, DOI 10.1007/978-3-642-10565-4_2, © Springer-Verlag Berlin Heidelberg 2011

Fig. 2.1 Ignition of a *Vulcain* engine on the launch pad during a *stage test* (Photo: CNES)

Fluids which are gaseous at ambient temperature but stored at low temperatures, below their boiling point, are called *cryogenic fluids* and rocket engines running on at least one cryogenic fluid (as fuel or oxidiser) are called *cryogenic engines* (Figs. 2.2 and 2.3). One of the first cryogenic rocket engines was used on the V2 rocket during World War II. The oxidiser of this engine was liquid oxygen (LOX).

LOX forms, together with liquid hydrogen (LH_2), a *high-level energy combination.* A typical specific impulse of such an engine is between 4,000 and 4,500 m/s. The *Pratt & Whitney RL-10,* developed in the 1950s was the first LOX/LH_2 engine used for spaceflight (Table 2.1).

Remark 2.1 *An aerospace scientist or engineer talking about the* ***engine cycle*** *normally means the cycle in which the engine produces its propellant (gas generator cycle, expander cycle, etc.), the* ***propulsion cycle****. In thermodynamics, the engine cycle is very close to the* ***Joule*** *cycle, and we could also talk about the life cycle of an engine type (design, production, test, flight service). The focus in this book is the* ***test cycle*** *or* ***operational cycle.*** *On the facility the engine goes several times through the cycle of pre-tests, final preparation, hot run and post-tests. On the launcher this cycle only happens once for an expendable engine.*

The special properties of a cryogenic fluid (see Appendix J) have to be considered within the operational aspects, particularly when it comes to preparation of a hot run. Vessels and lines need sufficient insulation, the feed system has to be free of pollution and a chill down phase is required before the hot run. The purging and venting of the rocket engine and the feed system starts immediately after shut-down and a cryogenic fluid must not be trapped in any of the line segments or engine components.

2.2 Test Facilities

To launch rockets just for test purposes or even in the development phase is history, at least for large launchers. A rocket engine dedicated for use on a launcher definitely needs its maturity for flight. The investment to lift a payload into orbit is in the region of some hundred million Euros and therefore, from first use on,

Table 2.1 Rocket engines running on LH_2/LOX as a fuel/oxidiser combination [8]

Engine	Manufacturer	Origin	Propulsion cycle	Thrust (kN)	Chamber pressure (bar)	Mass (kg)	Spec. impulse (Ns/kg)	Area ratio	Mixture ratio
Vulcain 2	SNECMA	France	Gas generator cycle	1,340	115	2,100	4,228	60	6.1
Vinci	SNECMA	France	Expander cycle	180	60	550	4,562	240	5.8
HM7B	SNECMA	France	Expander cycle	64.8	37	165	4,375	83.1	5
RL-10	Pratt & Whitney	USA	Expander cycle	66.7	20	135	4,248	40	5.9
SSME	Rocketdyne	USA	Staged combustion	2,094	207	3,393	4,434	69	6
LE-7A	JAXA	Japan	Staged combustion	1,096	121	1,700	4,316	52	5.9
RD-0120	KBChA	Russia	Staged combustion	1,962	219	3,450	4,464	87.5	6
RL-60	Rocketdyne	USA	Expander cycle	3,315	97	6,604	4,022	21.5	6

Fig. 2.2 Upper stage engine SNECMA HM7 (Photo: SNECMA)

Fig. 2.3 Space shuttle main engine, Rocketdyne (Photo: NASA)

a very high reliability of the propulsion system is mandatory. Unfortunately not all flight conditions can be simulated on a test facility and therefore a maiden flight will always remain a test flight. The hardware (engine dedicated for flight) and software (operational cycle and control programs) have to prove this high level of reliability on the test facility. On the facility the operational cycle is optimized, the engine is brought to maturity and during flight the engine finds its real application. In Europe's *Ariane* program an engine was normally tested twice, the second test being a precise rehearsal of the planned performance during flight. Before the first launch of an *Ariane 5*, its main engine *Vulcain* was tested 135 times on the facility *P*5 and about the same number of tests was performed on the almost identical facility *PF50* in Vernon, France.

The dimension and design of a test facility for rocket engines consequently follows the specimen and type of the requested tests. The size factor is easily dealt with, a small engine normally being tested on a small test rig, which is separated from the adjacent control desk by just a concrete wall. Large rocket engines are installed in large test facilities (Fig. 2.4), equipped with cranes, lifts, etc., and the control room is some 100 m away, in a bunker with several floors. The layout of the run tanks depends on the fuel/oxidiser combination. On a test facility for a cryogenic rocket engine, double wall, vacuum insulated tanks are used. The size of the tanks is matched to the consumption of the engine and to the requested test duration, the test duration again following the burn time of the engine during flight.

Before large rocket engines are tested in the completed configuration (*engine level test*), tests with single components of the engine are performed (*component level test*). The dimension of a facility for component testing can be large if we consider, e.g. that a test of a hydrogen turbo pump involves the same mass flow as on the complete rocket engine (Fig. 2.5). Hence the fuel supply has the same dimensions and the same safety systems are installed. The effort to set up the facility can be partially reduced if the demanded test duration is smaller than the flight duration or if the engine is not tested at full thrust.

The test facility has, besides the test bench and the control room, significant sub-facilities (Fig. 2.6). A very important sub-facility is the *altitude facility*. Engine tests under altitude conditions require a vacuum chamber, in which the engine is tested at low pressures equivalent to the pressures at high altitudes. Steam driven ejectors, powered by steam generators or steam vessels, are used to evacuate the vacuum chamber (Fig. 5.24).

Fig. 2.4 Test facility *P5* at German Aerospace Center (DLR) in Lampoldshausen for testing the *Vulcain* engine at sea level conditions (Photo: DLR)

Fig. 2.5 Component test facility for the LH_2 turbo pump of the *Vulcain* engine at SNECMA, Vernon [14] (Photo: SNECMA)

Fig. 2.6 Sub-facilities of the large test facility *P5* for the *Vulcain* engine at DLR, Lampoldshausen (Photo: DLR)

Chapter 3
Test Periods

Unlike during the flight, the rocket engine is normally tested more than once on the test facility. The *Vulcain* n° *M204* was tested 25 times on the facility *P5* and the accumulated test time totalled 13,788 s. Various tests (duration, operational points, regulation, etc.) are executed, covering the required range around the design point or around the operational point planned for the flight (see Sect. 4.5). Therefore on the test facility it is possible to perform extreme operational points, durations, various boundary conditions, and it is possible to test for malfunction of engine components.

In view of the test facility, the engine test represents an engineering process. It makes sense to operate the facility based on processes and to structure the facility into systems. We differentiate between systems directly used within the process and others. Further on, the operational team often has to receive instructions. The activities during pre- and post-test as well as the test execution are performed according to prescribed procedures and permanent approval of work (quality assurance) is installed.

3.1 Organisation and Clients of Test Programs

Within the development and flight phase of the European launchers, three types of test programs are applied. Development tests are executed first, followed by flight acceptance or qualification tests, and eventually the rocket engine is tested in the test program associated with the flight phase of the launcher. The European Space Agency (ESA) assigned the development of the *Ariane 5* to the French Space Agency (CNES) which again subcontracted the development of the various components of the complete launcher to European companies. Within the choice of suppliers, the principle of *industrial return* was applied. According to the financial contribution of the ESA member states, the orders were placed at suppliers of these countries. The commercialisation of the satellite launch and *Ariane 5* as a launcher was incumbent upon the company *Arianespace*. After the development was completed it was expected that the production and cost for the launch would be covered by the income from flight service. All tests of the *Vulcain* engine were ordered by the company SNECMA (formerly SEP), which was assigned for the

W. Kitsche, *Operation of a Cryogenic Rocket Engine*, Springer Aerospace Technology,
DOI 10.1007/978-3-642-10565-4_3,

development and final assembly of the engine. Within the development phase the administrator of SNECMA in turn was CNES and within the production/flight phase it was Arianespace.

3.1.1 Development Tests

During the development of a rocket engine several engines are produced only for the purpose of performing development tests. Accordingly, development campaigns are executed in which the settled objectives of the development are pursued. Many of the components and the operational cycle of the engine still undergo evolution in this phase. On the facility *P5* alone, 11 *Vulcain* engines were tested in 135 development tests before the first engine on this facility was qualified for flight 503.

Remark 3.1 *When Europe started the development of* **Ariane 5** *it had the high-flying target to transport someday the European shuttle* **Hermes** *(Fig. 3.1) into orbit. The men rated flight would also have demanded the utmost high reliability for the*

Fig. 3.1 Planning of the 1980s, the European shuttle *Hermes* on top of the launcher *Ariane 5* (Artist view) (Web site: ESA)

***Vulcain** engine. Such reliability has to be proved in a very high number of tests. Therefore, originally 500 development tests were planned on the two facilities **PF50** in Vernon and **P5** in Lampoldshausen.*

With the first tests the test cycle itself is developed, the result being the optimised operational cycle of the rocket engine. In the very first test a chill down of the engine and a short opening of a chamber valve is satisfactory. The following tests lead, e.g. from the first chamber ignition to the ignition of all pyrotechnical elements, to an operational point of 60% and eventually to the nominal operational point. The objectives of the following tests should be the test and optimisation of the engine regulation, the shut-down sequence and the exploration of the engine performance map. In parallel with these objectives, the durability of all components is demonstrated.

At the end of the development phase, failure tests are performed. Thereby the behaviour of the complete system is studied under the condition that one of the components has a malfunction. A typical failure test of the *Vulcain* is the non-activation of the chamber igniter. In this case the hydrogen/oxygen mixture is ignited by pilot burners (Fig. 5.19) which are in operation on the test facility as well as on the launch pad of the *Ariane 5* during every start up.

The further development from *Vulcain* to *Vulcain 2* (initially MK II) respectively concerned essentially the oxygen pump, its turbine and the divergent nozzle. In the new design the turbine exhaust was reintroduced into the main exhaust in the nozzle and thereby also used for film cooling of the inner surface of the nozzle (see Appendix D, Fig. D.1).

The *Vulcain 2* test campaigns started in May 1999, initially for a duration of 26 months with 84 tests and 5 engines planned. In fact, 80 tests were executed with 4 engines but the duration was 38 months. Six months later this engine failed during its first flight (see Remark 8.1) and supplementary development (see Sect. 6.4) was necessary.

3.1.2 Acceptance Tests

A rule of the *Ariane* program was to test each flight engine twice on the facility and, in exceptional cases, three times. The target to perform only one acceptance test was realised for the first time in January 2007 with the acceptance of a *Vulcain 2* engine on the facility *PF50* in Vernon. The stress of two acceptance tests never caused too much degradation of any engine component. Therefore none of the engines were blocked for use in flight. On the other hand, the sum of the chosen test durations (430 and 190 s) was representative of the operating time of 600 s during flight and the engine had demonstrated a second test cycle.

Remark 3.2 *The operational point of the **Vulcain** can be adjusted by a throttle in the supply line of the turbine on the oxygen side. On the test facility not only this throttle but also the valves at the inlet of the gas generator could be set to any desired* *position by means of hydraulic actuators. The resulting regulation mode made it possible to run several operational points during one test. In every first test of an engine acceptance this regulation was applied and so the individual characteristics of the engine were determined. According to the characteristics the engine could be adjusted exactly for the desired operational point for the planned flight. In the second acceptance test this operational point was demonstrated.*

3.1.3 Technology Support

After the development is finished and a first batch of engines of the serial production has passed their acceptance tests, normally a test program is started to support and accompany the technology applied during the flight phase (*Ariane Research and Technology Accompaniment*, ARTA) of the launcher. This test program is also applied to the engine, the object of these tests being to study and to clarify new technical questions occurring during the flight phase.

Remark 3.3 *After the first flights of the* ***Ariane 5*** *it was evident that the roll torque of the launcher was stronger than expected. The engines of the attitude control system (SCA) for compensation of this momentum were indeed able stop the rotation but consumed too much fuel. To study the problem the facility* ***P5*** *was equipped with a torque meter which was installed above the gimbals of the engine. It turned out that the helix like routed cooling channels induced a slight rotation of the exhaust jet. The perfect method to compensate the resulting torque was to adjust the outlet angle of the turbine exhausts.*

3.2 Test Campaign on the Test Bench

For testing purposes, for the duration of a test campaign the responsibility for the specimen is handed over to the operator of the test facility. The campaign starts with an official hand over meeting and is closed by the official return of the specimen to the client. The leader of the project is responsible for the overall execution of the campaign. The proceedings of the campaign are regulated by prescribed working procedures and controlled by the project leader or delegated to assistants responsible for the different areas of operation.

The scope of tasks during a test campaign is divided into:

- Handling of the specimen
- Compilation and management of work instructions (procedures) and computer programs (sequences)
- Test execution
- Analysis of test data
- Operation of the bench
- Operation of a measurement, control and command system (MCC)

3.2.1 Handling of the Specimen

The handling of the specimen is part of the pre- and post-test activities. It runs in parallel to the operation of the bench and the MCC system including the preparation of the test software. The permanent and mandatory coordination of these activities is the responsibility of the test leader.

Specimen handling involves:

- The transport and movement of the specimen
- The connection and disconnection of interfaces to the specimen, in particular the assembly and disassembly into (from) the test bench (Fig. 3.2)
- The inspections and checks on the specimen
- The shipping and reception of the specimen

In a broader sense, this handling involves:

- The coordination of the specimens handover, including the documentation, in particular the logbooks (Fig. 3.3) of the specimen and its components
- The analysis of failure reports concerning the test hardware
- Monitoring and maintenance of the work instructions (procedures) applicable to the specimen

The handling of the specimen has to be done according to the specification of the client (installation and operating manual) in parallel with the application of internal work instructions. The main objective within the engine handling is the transparent preparation for the hot run, the conclusion of the engine status after test and the controlled return of the engine to the customer.

Fig. 3.2 Integration of a *Vulcain* engine (Photo: DLR)

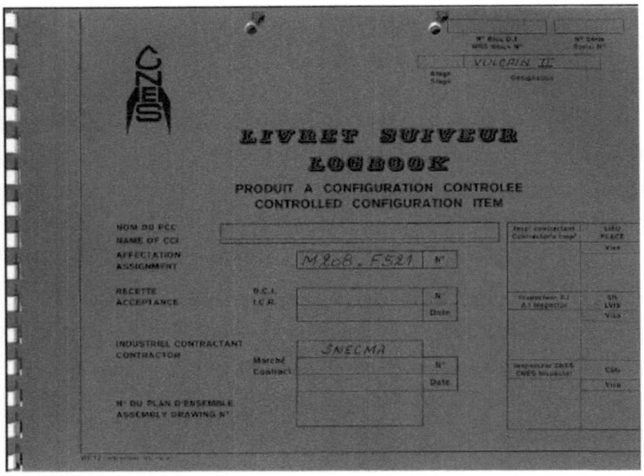

Fig. 3.3 Logbook of a *Vulcain 2* motor (Photo: CNES)

3.2.2 Work Instructions and Computer Programs

All information and data necessary for test execution constitute the *data base*. They exist in written form, on portable data media or they are stored in an online database and are part of the overall documentation. The project leader is responsible for the control of this documentation. The originator of this information/data is the customer (test request, sensor specification, inspection sheet, etc.) and the test team (bench measurement plan, specification of diagrams, bench adjustment parameters). The work instructions (*procedures*) and computer programs (*sequences*) refer to this data base and in turn they precisely and directly control the operation of the engine and the facility. Because of their direct influence on the test the procedures and sequences are subjected to an extensive validation process. The creation is in the hands of a member of the relevant work group, another member is responsible for checking the content of the procedure/sequence and another check against the applicable rules is performed by a quality assurance inspector. Procedures are presented to the user in a draft version and he adds his fair comments. The author uses these comments for a review, update and maintenance of the procedures. Sequences undergo a special validation before their application in a test. Due to the implicit risk, the value of specimen and facility as well as the high cost linked to a hot run, the sequences are partially tested three times. The test leader, the test requirer (client) and the quality assurance must all satisfy themselves that all sequences are in proper function before they enter the test.

3.2.3 Test Execution

The test leader plans and organises the test (test date, test team, meetings, fuelling, etc.), he controls the flow of information and the activities relevant for the test.

The activities on the test day (meeting, dry run, checks, final test preparation, hot run, handling of non conformance etc.) are the responsibility of the test leader. The results of the executed and evaluated dry run are presented to the test requirer in a *test readiness meeting*. All non-conformance reports relevant to the test have to be either accepted or closed and a risk analyse is closed as well. The status of the engine and of the bench systems is reconsidered and the test readiness is stated in a written form by the test leader and the test requirer. The scope of the various activities depends on the magnitude of the facility and the test. Directly after the test, in a post-test meeting, the test leader reviews the major events of the performed test and defines the next activities for the day after test.

3.2.4 Test Analysis

The test analysis by the test team covers the behaviour of the facility systems. It consists of an analysis of test data compared to previous tests, relative to the test request and relative to the specified performance of the facility. Detected deviations are reported to a non-conformance management process for treatment. A complete set of data is copied and stored by means of the facility. The test analysis concerning the test specimen is the task of the client, and therefore a structured set of test data is transferred to the client. He performs the analysis before the next test. The test team and test requirer treat in close cooperation those deviations which are linked to the interaction of specimen and facility. The test results concerning the facility are compiled by the test team in a *test report*.

3.2.5 Operation of the Test Bench

Any use of a facility system or equipment (cranes, movable platforms, analysers, etc.) belongs within the operation of the test facility. Mostly the operation means the performance of complex procedures such as a fuel transfer, a dry run, integration of components, preparation of tests and pre-tests. The operation is executed by the bench team on the basis of prescribed work instructions (procedures). Conducted procedures, compiled inspection sheets and non-conformance reports (NCR) form the feedback of the bench operation.

3.2.6 Measurement, Control and Command System

A team of specialists is responsible for the operation of the *measurement control and command* (MCC) system. The configuration of the MCC system is updated for the test in parallel with the integration of the test specimen. The basis for this update is the request by the customer concerning the parameters to be measured (measurement request) and the specified regulation and control. Further requests are defined by the test leader (e.g. bench measurement request). After analysis of the request a cabling plan is established; accordingly the measurement chains between sensors of the specimen and the data acquisition system are established and checked regularly. The hardware of the equipment (amplifiers, gauges, etc.) is adjusted and the control parameters and sensor characteristics are updated in the data base. The measured and archived data are finally saved (backed up) after test, converted into physical data, transferred to other computers (to the client) and printed in tables and diagrams.

3.3 Pre- and Post-Test Inspections

Before and after the hot run a number of inspections are standard and supplemented by inspections as a result of particular observations during the test. We discern functional tests (e.g. valve characteristics), leak measurement, visual inspections, boroscopic inspections, determination of dimensions, crack detection (dye penetrate test) and cleanliness checks (hygrometry, particle test) (Fig. 3.4).

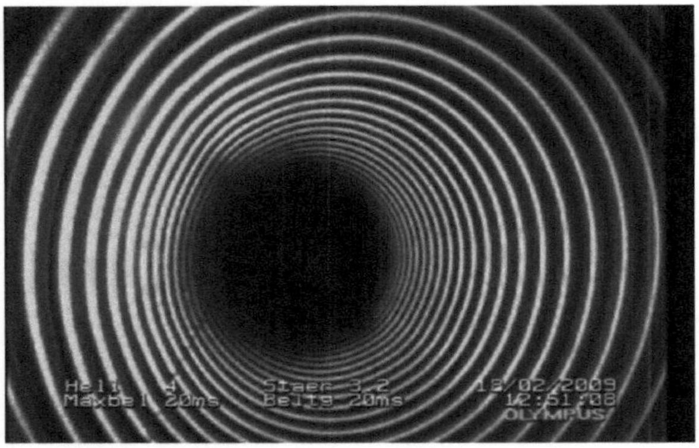

Fig. 3.4 Boroscopic inspection of a flexible tube (Photo: DLR)

3.3.1 Functional Test

Some of the subsystems of the rocket engine are tested regularly before each test. The dynamic seal of the *Vulcain* is such a subsystem (Fig. 3.5). The oxygen pump is directly coupled to the turbine which drives the pump. Because here the liquid oxygen is very close to the hot gas of the turbine, which has a high spill-over of hydrogen, the perfect separation of these two fluids is of the utmost importance. Therefore the casing of this turbo machine is pressurised between its two components with helium [2]. Any leakage of the sealings at the shaft only produces a helium flow. The function of this subsystem can be checked by pressurising a cavity and observing the pressure decrease vs time.

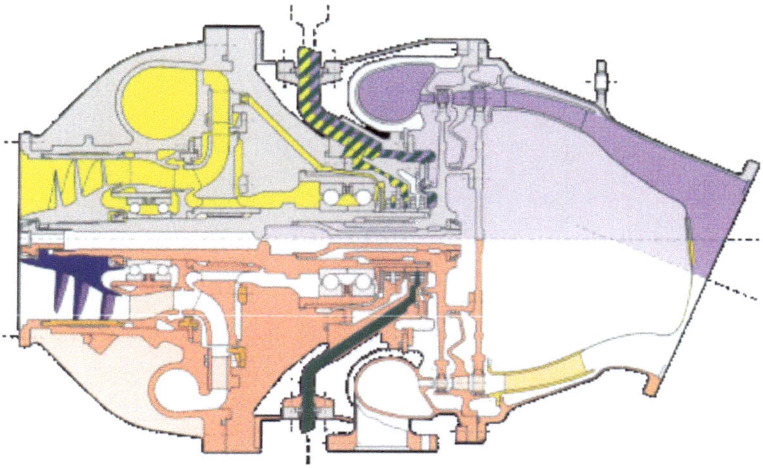

Fig. 3.5 Turbo pump and turbine of the *Vulcain 2* with dynamic seal between pump and turbine (Photo: SNECMA)

3.3.2 Leak Measurement

Leakage is an important component of anomalies occurring within the operation of rocket engines and launchers. It often causes delays and even malfunction. The terrible loss of the space shuttle *challenger* was caused by a leak. A systematic check of tightness of components is part of the pre- and post-test activities.

Leakage is usually checked by pressurisation with inert gas. An outward leakage of the engine or of a component of the facility can be detected by putting foaming fluids onto the potential points of leakage. Another method is pressurisation with helium and a check with a helium gas detector (Fig. 3.6).

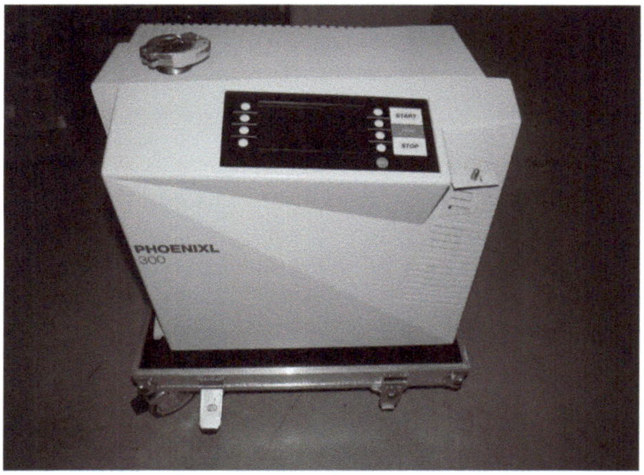

Fig. 3.6 Helium leak detector (Photo: DLR)

Fig. 3.7 Set of flowmeters (Photo: DLR)

Remark 3.4 *Experienced staff can not only to find a leakage by the generated sound but also estimate the leakage flow with sufficient precision.*

An internal leakage (e.g. across a valve seat) is checked by pressurisation from one side and observation of the pressure decrease or by measurement of the gas flow. If the volume of the pressurised cavity is known, the leakage flow can be calculated from the pressure decrease vs time. The leakage flow can be directly measured if a flow meter (bubble counter, buoyancy anemometer) is connected to the unpressurised side (Fig. 3.7).

Remark 3.5 *SNECMA, responsible for the final assembly of the* ***Vulcain****, developed a method to compute a leak flow in cold conditions and at high pressure (hot run conditions) from the measured leakage at ambient temperature and low pressure. The computed values were in very good accordance with available leak measurements taken during hot run.*

3.3.3 Visual Inspection

Visual inspections start with first access to the test cell after a hot run, being performed for the whole rocket engine and the whole test facility. Photos which are taken regularly before every test can be helpful to check any changes. Certain components (e.g. bellows, nozzle rim, stiffeners on the engine side, tube segments, exhaust guide system on the facility side) are systematically inspected. Visual inspections of inner parts are performed with boroscopes or endoscopes.

Remark 3.6 *After a **Vulcain** test the turbine starter was inspected and checked as to whether the burst disc which seals the powder had lost parts. The metallic parts of the burst disc partially floated through the hot gas line and the turbine. The parts induced the risk of a degradation of the turbine blades. Therefore after every test in which a loss of disc part was detected, an endoscopic inspection of the turbine was performed.*

3.3.4 Determination of Dimensions

A convergent and then divergent flow downstream the combustion chamber is typical for a rocket engine. This flow is realised by means of a convergent/divergent nozzle. The upper part of the *Vulcain* nozzle makes one structural unit with the combustion chamber, and the much bigger lower part, the nozzle extension (NE), makes a second structural unit. The enormous stress of the *Vulcain* NE causes a deformation at every hot run. Since 1993 the dimensions of the NE were measured after every hot run (Appendix C). Since 1999 the NE of the *Vulcain 2* was regularly surveyed using an optical/numerical method. The NE was tagged with marks, photographed and the pictures were analysed on a computer to study the three-dimensional deformation of the complete NE.

3.3.5 Pollution Control

Pollution is one of the undesired effects when a propulsion system is operated. We mostly speak about the pollution of the environment due to the exhaust from the propulsion systems. This pollution strongly depends on the fuel/oxidiser combination of the propulsion system. On the other hand the propulsion system itself can be very sensitive to pollution. For rocket engines running on liquid fuel/oxidiser, the operational aspects such as performance, function and reliability depend on the purity of the engine fluids. A basic requirement is that fuel, oxidiser and complementary fluids fulfil their specification. But the handling and operation of the rocket engine might also induce a risk of pollution. Pollution control [11] on a test facility for a cryogenic rocket engine means

- an accurate identification of sources of pollution,
- reasonable definition for pollution limits,
- application of adequate cleaning procedures, and
- systematic cleanliness checks.

The structure, scope and details of pollution control procedures depend mainly on the test specimen and on the design of the test facility.

3.3.5.1 Sources of Pollution

When a cryogenic rocket engine is operated, three main kinds of pollution have to be considered: particles, foreign gas and humidity. The first potential source of particle pollution is the rocket engine itself. Several procedures within the production process can create deposits inside the components of the engine. There is the risk of metallic parts due to the milling process and the risk of weld beads and cinder due to welding. The same pollution is possible for many facility components (Fig. 3.8). The requested cleanliness level has to be considered within the definition of a system, because the choice of material and the fabrication process limits the possible cleanliness level of many installations. A proper cleaning procedure before integration can reduce this kind of pollution. During operation of the rocket engine and the facility, further particle pollution is possible. All supply lines induce the risk of particle pollution. A good filter configuration and filter management can reduce the introduction of foreign particles. The operation also induces a risk of internally created particles. Due to friction, the combustion process, the running of pyrotechnical elements and corrosion, different kinds of pollutions have to be faced. This kind of pollution either has to be accepted or has to be cleared after every engine operation. The polishing of the inner surface of a combustion chamber is an example of such clearing of this kind of pollution.

Filters normally prevent particle introduction but not the introduction of humidity and air. The presence of both is accepted during integration and cleaning from humidity and foreign gas is performed when a fluid circuit is mechanically closed. Nevertheless there are several chances for the reappearance of this kind of pollution during operation. The facility and the engine have venting openings which cannot

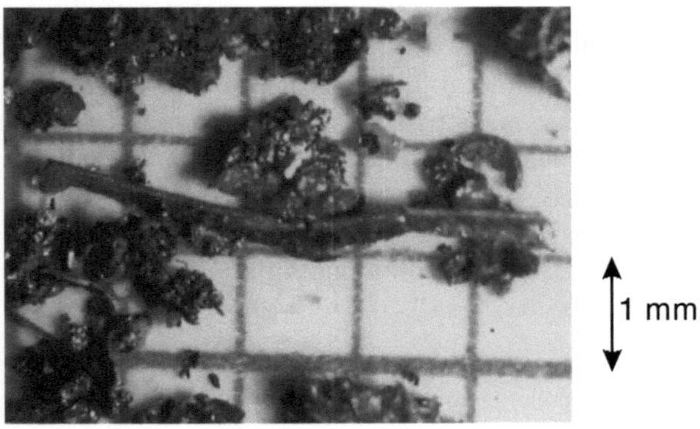

Fig. 3.8 Initial particle sample before cleaning and integration of a subsystem in a storage area (Photo: DLR)

be perfectly tight. Also the internal connections can cause pollution by fluid from one circuit travelling to a circuit of another fluid.

An example of internal pollution with gas was studied on the oxygen supply of the facility *P5*. Most of the oxidiser vessels of a launcher or a test facility have a pressurisation system (see Fig. 6.2) to push the fluid towards the rocket engine. The pressurisation of liquid oxygen (O_2) with gaseous oxygen would be the best choice from the pollution aspect. But for safety reasons, gaseous nitrogen (N_2) is often used instead of oxygen. In this case the condensation of N_2 at the surface of the liquid O_2 has to be considered (see Appendix H).

3.3.5.2 Sensitivity of the Facility and Engine to Pollution

Any pollution of the fuel or oxidiser can have an impact on the performance of the propulsion system. The specific impulse, the combustion pressure or the temperature can change when, e.g. the liquid oxygen is polluted by a certain amount of nitrogen. This kind of pollution is normally very small and so is the impact on performance.

Pollution with humidity is much more important because it concerns not only the performance but rather the functionality and reliability aspects of the rocket engine. Water in liquid oxygen would instantly turn into ice particles and is therefore as severe as any other particle pollution. In the different components of the rocket engine and the test facility the fluids have to pass through not only pipes of large diameter but also tiny ducts and slots. On the facility the filters can be blocked by particles and on the engine the particles can be captured in flow cooled bearings and sealings, in the injection elements (Fig. 3.9) of a combustion chamber or in cooling channels. Even a small amount of pollution of this kind can have a drastic impact on the function and reliability of the rocket engine. On the hydrogen side the conditions are even more important because any foreign gas (except helium) will also turn into ice when it comes into contact with liquid hydrogen.

The particle pollution of the *Vulcain* chamber injectors was studied in a special campaign on the facility *P5*. Injector elements in selected circumferential and radial positions were partially, artificially blocked and the impact on the injector plate was checked by an inspection of the surface after test (Fig. 3.10).

The blocking of filters due to ice particles is hard to identify because an inspection at ambient temperature cannot reveal the pollution. An ice blocking of a small sensor tube in the vicinity of the hydrogen vessel of the facility *P5* caused an important reduction of the reliability of the facility. The sensor concerned was one of three sensors used within the inlet pressure regulation for the rocket engine. Blocking of the tube was avoided by an adequate venting procedure.

Oxygen pollution in liquid hydrogen [7] not only concerns the functional aspects but even more, and more important, the safety aspects of the facility. Oxygen ice particles can react spontaneously in liquid hydrogen and even a small fraction can cause a pressure increase in a vessel. Pollution with chemically reacting substances is yet more dangerous on the oxygen side of the test facility. An ignition inside

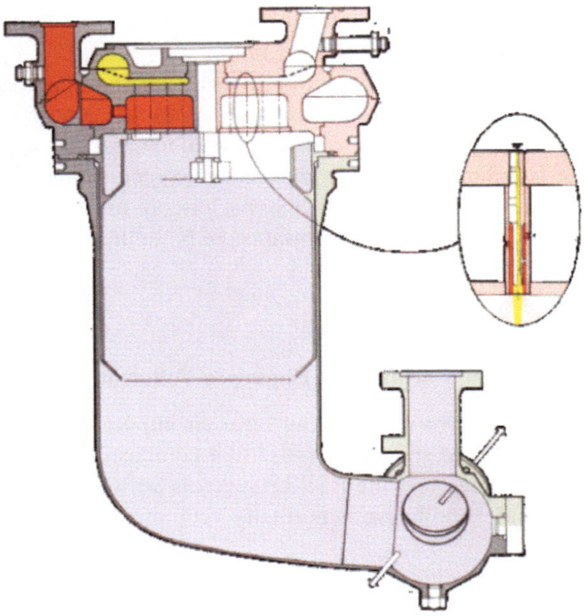

Fig. 3.9 *Vulcain* gas generator with injection elements (Photo: SNECMA)

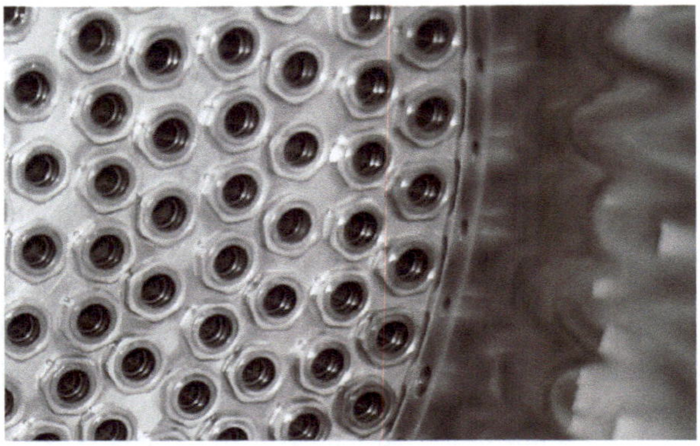

Fig. 3.10 Visual check of a thrust chamber injector plate (Photo: DLR)

liquid oxygen can start the burning of the surrounding metal, i.e. the pipe or the vessel itself.

Pollution of secondary fluids can affect the reliability of the facility. Pilot valves, for example, cannot work properly if the pneumatic gas has high particle pollution (see Fig. 3.8). Pilot valves are furthermore sensitive to pollution caused by insects, which intrude by way of the outlet into the atmosphere and block the small tubes.

3.3.5.3 Cleaning Procedures and Checks

All vessels, lines, valves and any other component of the flow circuit of a test facility have to be cleaned before integration. The total inner surface of the hydrogen tank (600 m^3) of the *P5* and *PF50* are washed with a dissolver to get it free of particles and grease. Lines are etched with acid, rinsed with demineralised water and sometimes put into an ultrasonic bath (Fig. 3.11). Directly before integration the preassembled segments of the fluid circuits are cleaned with inert gas. Normally pure nitrogen is used to blow through the system. After integration the replacement of nitrogen by another gas is performed by a so-called *conditioning procedure*. A dedicated number of cycles of pressurisation and depressurisation are performed on the segment in order to reach the required purity of the final gas filling.

The best protection against pollution after installation is the fact that the circuits are closed, are equipped with filters and maintained at a slight overpressure.

For the interfaces between the test facility and the rocket engine, cleanliness concerning particles is obtained and proved during the integration of the engine into the test facility. For that purpose the connecting lines between engine and facility are blown through with nitrogen and the remaining particles are collected in a filter. This filter acts as a particle collector and a microscopic examination reveals the size and number of particles in the sample. Table 3.1 gives the criteria for particle cleanliness for the *Vulcain* engine.

The purity concerning foreign gas and humidity has to be established and proved before every engine test. Normally helium is used on the hydrogen side and nitrogen on the oxygen side of the test facility to purify the lines. During the cycles of the conditioning procedure the polluting fraction is diluted again and again until the requested cleanliness criteria (Table 3.2) are reached. At the end of the procedure samples are taken in small bottles and the content is analysed with a gas chromatograph.

Fig. 3.11 Ultrasonic cleaning station (Photo: DLR)

Table 3.1 Particle cleanliness requirements of the *Vulcain*

Circuit	Pneumatic actuators	Fuel/oxidiser lines	Hot gas lines
Size	Max number		
15–25 μm	1000	2000	4000
25–50 μm	100	200	400
50–100 μm	10	20	40
100–500 μm	1	2	4
> 500 μm	0	0	0

Table 3.2 Cleanliness criteria concerning gas and humidity for the hot run of the *Vulcain* engine. Water (humidity) and any other gas except helium would turn into ice particles in liquid hydrogen, which are as harmful and unacceptable as any other rigid particles

Flushed area	Gas	Max. allowed humidity	Max. allowed fraction of polluting gas
Feed line liquid oxygen	Nitrogen	< 15 ppm	No limitation
Turbo pump liquid oxygen	Nitrogen	< 15 ppm	No limitation
Feed line liquid hydrogen	Helium	< 15 ppm	No limitation
Turbo pump liquid hydrogen	Helium	< 15 ppm	< 1000 ppm nitrogen, < 100 ppm oxygen
Hot gas circuit	Helium	< 15 ppm	No limitation

Another check of pollution is performed online during the final test preparation and chill down of the cryogenic engine. This is, at the same time, a check of internal and external tightness. The close surrounding of the rocket engine is monitored with two gas analyser systems, each working on a different physical principle. Inside the rocket engine further gas concentration measurements are taken. They can reveal a possible internal leak during the chill down phase. The gas samples are taken with *Teflon* hoses which burn off after ignition.

Remark 3.7 *During the erection phase the particle cleanliness of the fluid circuits on the test facility* ***P5*** *was monitored. It turned out that the manufacturing process for the gas tubes was not compatible with the relative high cleanliness required. The tubes were unexceptionally made of stainless steel and were pickled and rinsed after fabrication and mounting in assembly groups. The inner surface of the tubes was still tainted with unacceptable particles which could not be rinsed but could be separated by application of shocks, e.g. by light hammer strokes on the lines.*

Chapter 4
Engine Test

An important character of a rocket engine and its components is the lifetime. Because we don't have plenty of it. As well as the lifetime, the number of engine cycles (hot runs) is limited. Therefore the consumed lifetime and cycles performed are subjects of exact accounting in the engine and component logbooks. For this purpose, a precise definition of what counts as a test is necessary. Concerning the contractual aspects between the operator of the facility and the customer, we speak about a test as soon as the feed lines of the test facility are filled. For this step the test clearance is necessary which means the complete test preparation is successfully performed. Even in the case of abortion shortly after this step, from the contractual view the test is considered as performed.

Within the technical aspects a test is counted if the chill down of the engine had started; from that point on the engine has to face burdens which are only encountered during operation and, respectively, during a test (e.g. valve manoeuvring, thermal load). As soon as an ignition had been effected the test is taken as a *hot run* (Fig. 4.1).

4.1 Test Configuration

Both the rocket engine and the test facility, including the MCC system and its data base, have to be set into the defined test configuration before a test can proceed. The definition of the test configuration is on the test requirer side. In the *test request* (see Sect. 10.2.1) he defines which components of the specimen are to be exchanged, and which to be added to or removed from the specimen He defines the pressure, temperature and flow at the interfaces between rocket engine and test facility. Normally the mounting of the pyrotechnical elements is the last mechanical work before the test. All components of the rocket engine are monitored by quality assurance according to their logbooks which contain the calculated life duration of the components. None of them must be used in a test after its specified life duration has ended.

To reach and to confirm the test configuration the test team performs several procedures in parallel. The inspections and pre-/post-checks have to confirm that no unacceptable anomaly has occurred with the specimen. The data base containing the information regarding in which manner the acquisition and archiving of the

W. Kitsche, *Operation of a Cryogenic Rocket Engine*, Springer Aerospace Technology,
DOI 10.1007/978-3-642-10565-4_4, © Springer-Verlag Berlin Heidelberg 2011

Fig. 4.1 Hot run of a *Vulcain* rocket engine (Photo: DLR)

requested measurements is to be done and the sequences for control and regulation are updated in parallel with the mechanical procedures. The run tanks and gas bottles are also filled to a level which fits the requested test duration and pressure profiles. In case of substantial changes of the hardware or software configuration the total configuration is validated by a preliminary dry run; afterwards, if necessary, final corrections are done. For a dry run, as for a hot run, all interfaces between the facility and the engine are adjusted according the test request and all sequences are activated in a dry run as if we were in a real hot run (Table 4.1).

Table 4.1 Fluid interfaces of the *Vulcain* engine to the facility *P5*

Number	Fluid
2	Liquid hydrogen
2	Liquid oxygen
1	Liquid helium
7	Gaseous hydrogen
29	Gaseous helium
13	Gaseous nitrogen
10	Hydraulic oil

4.2 Test Readiness

The final criteria for test readiness are laid down in the test request. They are checked and finalised on the testing day. For that purpose the closures and venting points are opened, a final dry run is performed and the feed lines as well as smaller lines and inner cavities of the engine are flushed with inert gas. After this dry run no change of the configuration (hardware or software) is allowed; the hot run has to be executed in the same configuration as the last dry run before the test readiness statement. The correct function of valves and activated systems is checked by means of the data recorded during this dry run. The cleanliness (see Sect. 3.2.6) of the tubes and the engine after flushing is confirmed by analysing samples of gas which are taken from various sections of the system. Finally, in parallel with these checks, the document which confirms the successful performance of all mechanical pre-test activities and that the requested test configuration is reached is produced. Any unresolved deviations detected during the test preparation (see Sect. 8.2) now have to enter the assessment, in which the decision has to be taken whether or not the hot run can be performed despite the deviations.

4.3 Hot Run Preparation

To ignite a rocket engine in order to start a normal hot run, special procedures are necessary which transfer the engine from the rest status into the condition ready for ignition. All chemical propulsion systems require filling of the feed lines in this phase and defined pressures in front of the chamber valves have to be adjusted. For rocket engines running on cryogenic fluids, the chill down of the engine and its feed lines is of particular importance. Further on, special procedures may be necessary due to the design or due to the cycle of the engine or its components.

4.3.1 Chill Down of Cryogenic Rocket Engines

A propulsion system running on cryogenic fluids has to pass a chill down phase before ignition. This chill down is indispensable, particularly if the propulsion system has turbo pumps to deliver the extremely cold fluids. The feed lines, turbo pumps, valves and ducts of the engine have to be brought down to the temperature of the cryogenic fluid. During the hot run the turbo pump delivers the fluid at high rotational speed. Starting the pump without prior chill down would cause cavitation, the bearings would not have sufficient cooling (see Remark 4.1) and the fuel or oxidiser would be pumped into the chamber with an unacceptably high fraction of bubbles and gas.

The necessity for a chill down is already considered in the design phase of the engine. The *cryo circuits* of the engine are therefore equipped with valves which

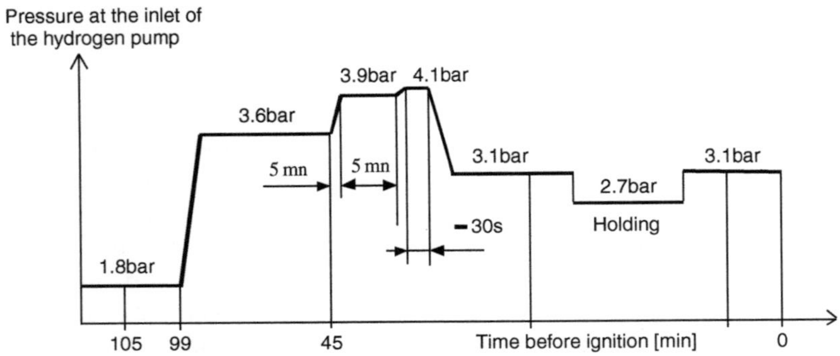

Fig. 4.2 Typical chill down profile of a *Vulcain* engine (Test Request: SNECMA)

enable the cryogenic fluids to be returned at convenient points back to the bench lines. On the facility side the fluids are recycled to the process or disposed of in a suitable manner.

The duration and pressure profile (Fig. 4.2) of the chill down depends on the conditions of the rocket launch procedure. Both are chosen with enough margins so that the chill down criteria (Table 4.2) are matched at the end of the phase.

Remark 4.1 *During the* ***Ariane*** *flight L5 one of the bearings of the oxygen pump of the HM7B engine of the third stage was not cooled sufficiently. At 160 K the pump finally overheated, exploded after 560 s flight time and the mission failed. Later on the investigation board revealed that in previous flights the pump only worked on the back up cooling.*

Table 4.2 Chill down criteria on a *Vulcain* engine

Measurement point	Requested temperature
Inlet of the hydrogen pump	20 K < T < 22.5 K
Outlet of the hydrogen pump	20 K < T < 30 K
Bearings of the hydrogen pump	17 K < T < 45 K
Inlet of the oxygen pump	90.1 K < T < 92.5 K
Outlet of the oxygen pump	90.1 K < T < 93.1 K
Bearings of the oxygen pump	87.1 K < T < 100 K

4.4 Hot Run

The operational cycle (see Remark 2.1) of a rocket engine depends on the design and the propulsion cycle of the engine. The *Vulcain* is a bipropellant engine working in the gas generator propulsion cycle (see Appendix F, Fig. F.1). The hot run as part of the operational cycle begins, by definition, at the moment the command to open the chamber valve for hydrogen is given. Although the actual ignition happens a bit later, the MCC system marks the release of this command with a time stamp which is recorded and understood as the ignition time H_0. The hot run ends when the shut-down sequence is called. Its first command is to close the valve for the oxygen supply of the hot gas generator.

4.4.1 Start Up Transient

The principle course of the start up transient is already fixed in the development phase before the construction. The design of the components which are activated to perform the start up (pyrotechnical elements, venting/purging valves, chamber and gas generator valves) fits the need during start up (burning time of pyrotechnical elements, valve characteristic). The principle order of activation is also settled at an early state. For the *Vulcain* the development company decided to ignite first of all the chamber [2]. For that purpose the chamber valves are opened, first for hydrogen, then for oxygen, and the chamber igniter is activated. The chamber burns now at a low pressure which complies with the tank pressurisation. With a closer look, the mixture ratio (O_2/H_2) levels out in this phase and the ratio is locally and temporarily highly inhomogeneous. The mixture ratio is partially far above (up to m_{O2}/m_{H2} equal to 20) the stoichiometric point. To perform this phase at a low combustion pressure reduces the stress on the inner chamber wall. A venting system (He gas, 10 g/s), which is mainly used for the cleaning before and after the test, is activated during this phase to support and expedite the mixing process and to avoid back flow.

The next step is to start the turbo pumps. The gas generator to feed the turbines of the pumps delivers only sufficient flow if its inlet pressure is sufficiently high. Therefore the pumps have to be started by another component. For that purpose, a special turbo pump starter is used, a pyrotechnical component with a long burning time. The pyrotechnically created hot gas is determined in such a way that the pumps can spin up to 60% of their nominal rotational speed before the hot gas supply is switched over to the gas generator. To match this switch over, the gas generator is ignited by the third pyrotechnical component (gas generator igniter). Its ignition is performed at higher pressure as for the chamber but it is running at significantly lower mixture ratios than the chamber.

The order of valve and igniter activation for the chamber was fixed at the outset. But for the gas generator different orders of activation (Fig. 4.3) for the turbo pump starter, gas generator igniter and gas generator valves were studied during the first test campaigns of the development period. Both the order and the timing of the activation signals were tested, and also for the chamber components.

The activation time has to be adjusted exactly to the characteristics of the different components. A valve has a characteristic of movement (Appendix A), the characteristic of a pyrotechnical component is the profile of pressure vs time (Appendix B) and the characteristics of the turbo machines are reflected in performance maps (see Sect. 4.5).

Remark 4.2 *The activation commands from the MCC have a specific delay (see Table 4.3). It depends mainly on the load of the MCC processors and for the original*

Fig. 4.3 Start up sequence of a *Vulcain* (Chamber ignition, start of pumps, full flow) (Image Video: DLR)

Table 4.3 Examples of delays for commands of a start up sequence

Command	Requested activation time [s] Relative to the release of the start up sequence	Delay [ms]
Purge valve closing	0	1
Hydrogen chamber valve opening	1.00	8
Oxygen chamber valve opening	1.55	8

***P5** MCC system it was sometimes higher than 10 ms. When the MCC was replaced in 2008 these delays were barely visible any more in the valve pre-checks. Delays were now given in μs.*

4.4.2 Control and Regulation

In principle we have two modes of engine control, the unconditioned control (*open loop*) and the regulation of target parameters (*closed loop*). Target parameters are the mass flow and the mixture ratio (m_{O2}/m_{H2}) of the propellants, the combustion temperatures and the thrust of the rocket engine. Some of the parameters are directly measured, others are computed from measurable parameters. In the open loop mode the valves are switched to defined positions and the engine adopts its operational point according to the individual engine characteristic. In the closed loop mode one or more target parameters are defined and the control valves are adjusted according to an algorithm until the measured values equal them. Normally other conditions beside the valves position have an influence on the target parameter and on the operational point. The chamber pressure of the *Vulcain* is basically a function of three valve positions, the two gas generator valves and a throttle in front of the turbine on the oxygen side. Further influence comes from the pressures and temperatures at the turbo pumps inlets. At small changes of the control valves the first-order approximation of the operational point is taken as linear. The linear relation can be assumed for all target parameters to all control parameters and hence the inversion is also linear. Hence we can compute within the running loop the correction of the control parameters from the deviation of the target parameters from the measured values. The coefficients of the linear relation turn out as a matrix which again depends on the actual operational point. Theoretically the knowledge of the matrix at any point is necessary to perform a proper regulation. But it is the principle of a closed loop regulation to measure permanently (in *real time*) the operational point and to correct it by means of the control elements in case of a deviation. Therefore a regulation matrix can be used in which the coefficients are constant averaged values from the total region of the performance map.

The typical duration of a regulation cycle is 0.2 s, and within the cycle 125 samples per second are taken.

The regulation specification of the *Vulcain* is a volume of 38 pages and, besides the described principle of the regulation, it mainly contains the rules on how to start and to stop the regulation, how to validate the measured values, how to correct the measured and target values, how to compute the target parameters from measured parameters and the numerical values of the coefficients in the regulation matrix. The sampling rate (see Remark 4.3) of the measured parameters also has an influence on the performance of the regulation and has to be well defined.

Remark 4.3 *The importance of the correct sampling rate was revealed in a regulation failure during a* ***Vulcain*** *development test. The sampling rate of the temperature sensors was too low relative to the regulation cycle. Normally temperatures are acquired at a relative low rate (e.g. 10 Hz); a higher rate is not necessary because the measured temperature variation in a rocket engine is not so fast that higher sampling rates are required. But the design of the regulation algorithm required a rate of 125 Hz for all parameters used. The absent temperature values caused a malfunction of the regulation. The position of the valves started to oscillate in a period of 4 s at an amplitude of five degrees around a mean value. The oscillation of the target parameters had the same frequency and the chamber pressure oscillated between 100 and 107 bar. The amplitude of the gas generator temperature was even 125 K. Despite the high amplitudes, all parameters remained within the monitored limits and therefore no redline was violated to stop the test.*

4.4.3 Monitoring of Engine Parameters

Several important engine parameters are permanently monitored during the hot run. The concerned parameter is accepted within programmed limits. In case of a limit violation a programmed action of the MCC system is released. This action can be an alert, the switch off of the regulation, the switch on of a venting and it can also be the release of the shut-down sequence (Fig. 4.4).

The monitored parameter is almost never derived from only one measured value. Otherwise a sensor failure could initiate an action, even an abort of the test although no real deviation of the parameter is existent. On the *Vulcain*, as an example, the hot gas temperature is derived from four adjacent measurement points. Only if three of them are out of the range (300–1200 K) is the test aborted. If we consider the case of a sensor failure the importance of a lower limit becomes evident. A defect sensor could provide a zero value even when the real value is above the upper limit.

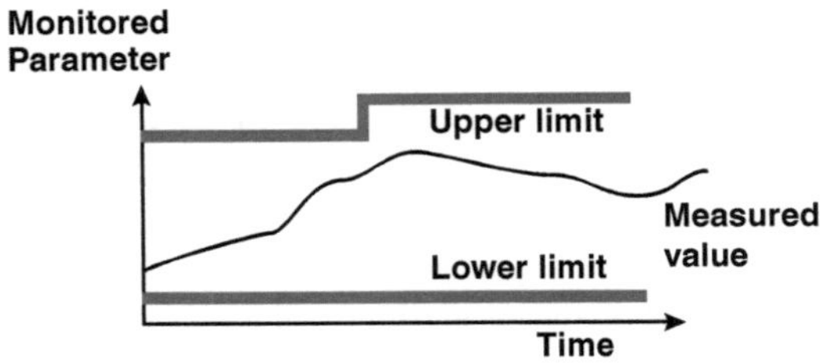

Fig. 4.4 Upper and lower limit *(redline)* for the monitoring of a parameter (Photo: DLR)

That means that even when all four real values are above the upper limit while two sensors are defect (and show zero) no test abortion would be initiated in the example above.

The logic to continue a test as long as n out of m parameters are within the limits is called a *positive majority logic*. It is also a matter of definition to use a *negative majority logic* where a test is stopped when n out of m parameters are outside the limits. For the monitoring of a shaft speed normally a positive logic 1/2 (one out of two) is applied, for vibrations 1/3 or 2/4 (Table 4.4).

Table 4.4 Typical shaft speed monitoring for turbo pumps

		Majority logic (pos)	Lower limit [1/min]	Upper limit [1/min]
During start up	Oxygen pump	2/2		16, 000
	Hydrogen pump	2/2		37, 500
After start up	Oxygen pump	1/2	4, 000	16, 000
	Hydrogen pump	1/2	4, 000	37, 500

The check of a parameter at a defined time (no permanent monitoring) is called a *punctual check*. The punctual check is embedded in the same philosophy of majority logic as the redlines.

Remark 4.4 *During the first test campaign on the facility* ***P5*** *more and more monitorings were installed. At that time the test team did not know that the number of monitorings on the MCC system was limited to a relatively low number. When the number was exceeded the MCC system did not start* ***any*** *monitoring and on top of this no warning was produced. One of the parameters in the assumed monitoring was the chamber pressure at 120 bar. Due to the uneffective monitoring the engine reached 122 bar in the seventh test and after early concern the persons in charge were happy about the high performance.*

4.4.4 Shut-Down Transient

Analogous to the definition of the ignition the shut-down of the engine is defined as the moment when the first signal of the shut-down sequence is sent off by the MCC system. The first command of this sequence is the closing of the oxygen valve

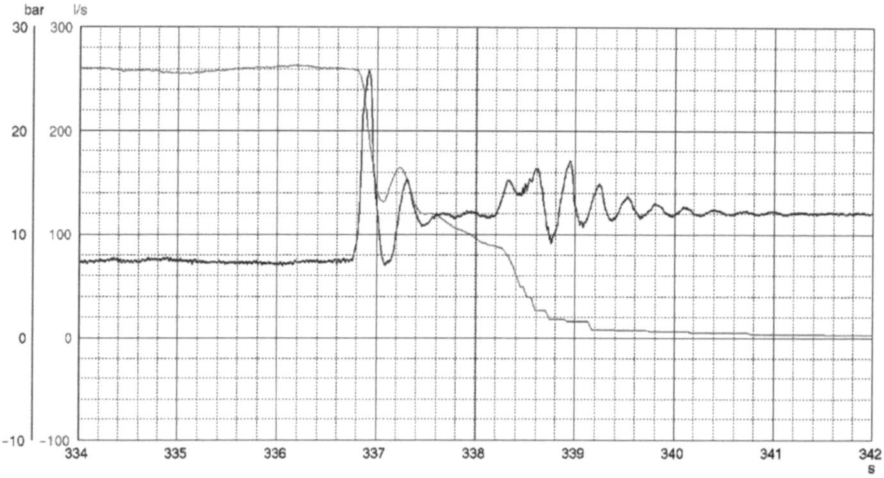

Fig. 4.5 Feed line pressure and oxidiser flow during a water hammer in an oxygen feed line (Photo: DLR)

of the gas generator and hence its shut-down. This stop effects the reduction of turbine speed and a pressure drop in the chamber feed lines. In principle the order of the shut-down sequence is the inversion of the start up sequence. The last action is to close the chamber valve for liquid hydrogen. During shut-down the sudden deceleration of the fluids in the lines has to be considered. The closing times of the bench valves in the feed lines are therefore adjusted to assure that no water hammer (Fig. 4.5) occurs but, on the other hand, the close position is reached as soon as possible.

The feed lines have venting points and access for flushing gas to flush – in parallel with the shut-down – all separate sections of the lines. The venting points of the engine are opened as well and the flushing is activated. The flow in the chamber and nozzle is very unsteady during shut-down, pressure temperature and mass flow rapidly decrease and the nozzle flow turns from supersonic to subsonic flow. Normally we observe a continuous decrease of the parameters but, due to an inappropriate timing of the valve closing, an oscillation (*chugging*) [4, 17] of the chamber pressure can occur (Fig. 4.6).

During shut-down the pressures before the injection elements drop quickly below the critical pressures of the cryogenic fluids. On the *Vulcain* we observed chugging at pressures below 13 bar ($P_{krit,H2} = 13.16$ bar); below this level gas bubbles can emerge in the liquid. The fluid, now in two phases, becomes sensitive to impulses which cause an oscillation and which are multiply present.

For the operation of an expander cycle engine the chugging effect has a particular importance. Normally the chamber pressure in the nominal point of an expander cycle engine is higher than the critical pressure of oxygen (50.9 bar) but mostly there is also a request for low operational points, which means lower pressure. In this case there is a risk of oscillation in the oxygen feed line of the chamber which can flash over to the combustion in the chamber.

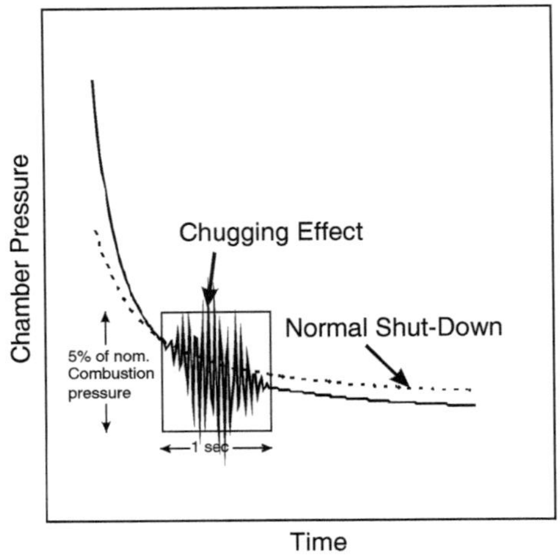

Fig. 4.6 Oscillation of the chamber pressure during shut-down (Photo: DLR)

The chugging effect was the subject of one of the test campaigns with the *Vulcain*, but no negative impact on the engine was found.

4.5 Engine Performance Map

Just like in all technical areas, one is no longer satisfied with only one value (e.g. the thrust) to characterise a rocket engine. Depending on the adjusted mixture ratio (say at constant mass flow) the thrust comes out as a characteristic curve. If another parameter is varied we obtain a set of curves and call it a *performance map*.

The rocket engine performance map is used to describe the operational behaviour of the engine in steady state. The performance map of the complete propulsion system could be a diagram which shows the thrust vs the mixture ratio ($\dot{m}_{LOX}/\dot{m}_{LH2}$) (Fig. 4.7).

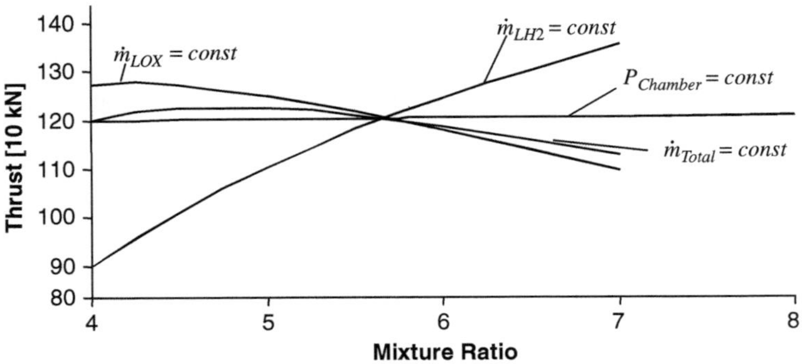

Fig. 4.7 Performance map of a hydrogen/oxygen rocket engine, designed for 1200 kN thrust (Chamber Parameters) (Photo: DLR)

The performance map can be used to show the reference point, curves of constant combustion pressure, of constant mass flow and to show a curve of regulation, the curve from one operational point to another. Further on the performance map has to show the limits of operation. The operation of a combustion chamber, the heart of a rocket engine, is limited due to a maximum pressure and a maximum mixture ratio (Fig. 4.8).

Performance maps are also used to illustrate other components of the rocket engine. Figure 4.9 shows the performance map of a turbo pump.

For the turbo machine designer, the behaviour of the chamber and, respectively, the injection elements is the *characteristic of the load*. Normally the pump flow is not equal to the chamber flow (if a gas generator is supplied in parallel), but if the flow ratio is known the characteristic of the load can be introduced into the performance map of the pump (Fig. 4.10).

One of the operational limits of the pump is the maximum shaft speed. A look into its performance map gives us the maximum flow as a function of the requested chamber pressure, which means another operational limit.

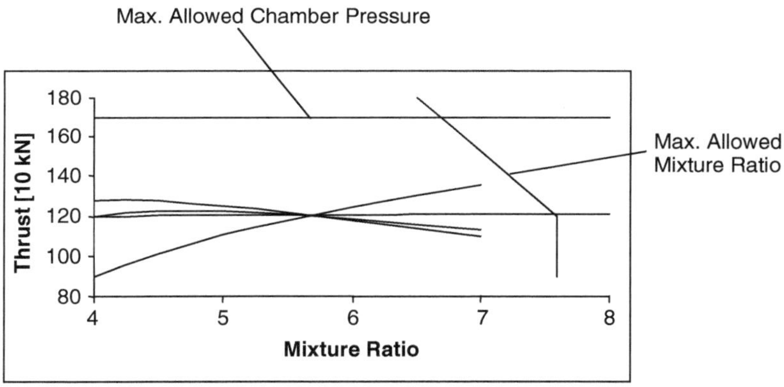

Fig. 4.8 Limits in the performance map of a rocket engine (Photo: DLR)

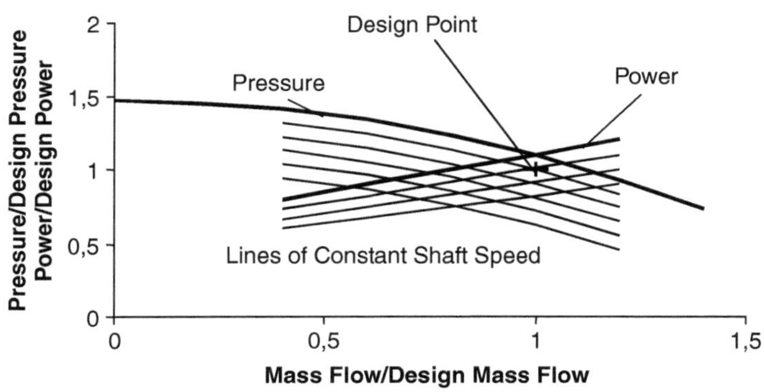

Fig. 4.9 Performance map of a turbo pump (Photo: DLR)

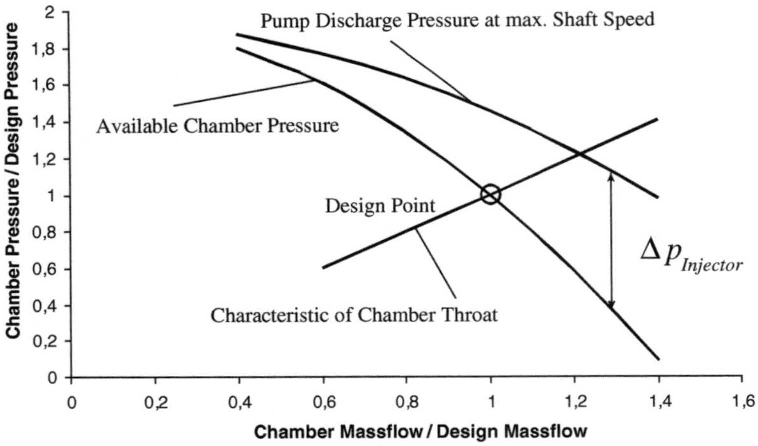

Fig. 4.10 Linked pump and chamber characteristics (Photo: DLR)

We can conclude that the operational points of the different components are linked and that their operational limits are a result of an overlay of the different performance maps (Fig. 4.11).

Remark 4.5 *Shortly after the start of the* ***Vulcain 2*** *development the test program was in a dilemma. The oxygen pump developer* ***reduced*** *the maximum chamber pressure temporarily because, for the time being, the maximum rotational speed was not yet allowed. And the nozzle developer* ***increased*** *the minimum chamber pressure because the risk of back flow and extra combustion at the nozzle exit was not yet clarified. The extra combustion would induce unacceptable high temperatures at*

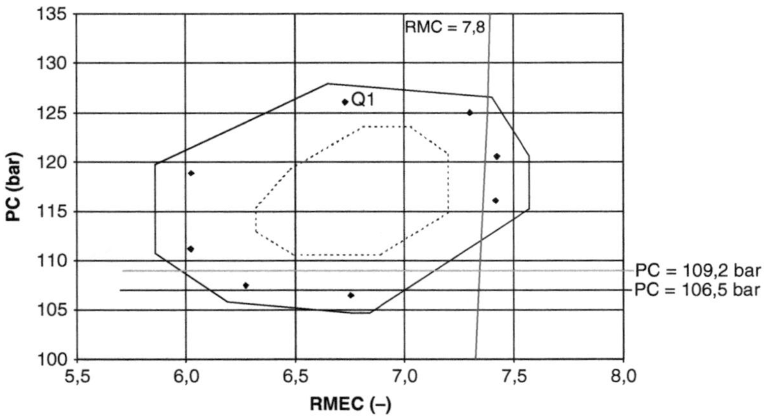

Fig. 4.11 Performance map of a *Vulcain* chamber (Test Request: SNECMA)

the nozzle rim. Within these conditions the test engineers had to work and had only a small section of the performance map available.

The nonlinear and only partially analytical given relations between the engine parameters require a numerical computation of the performance maps. Their permanent validation and correction is an essential requirement during the test period. Performance maps have a central importance within the operational aspects of a rocket engine; they are at the centre of the dialog between test engineer, chief engineer and all designers of subcomponents of the propulsion system.

4.6 Test of Expander Cycle Engines

Upper stage engines designed for an expander cycle are qualified in an *altitude facility* (see Sect. 5.8). The test including the chill down is performed in a vacuum chamber. After shut-down a second chill down and the demonstration of *reignitability* can follow (see Fig. 4.12) [9].

Normally the expander engine as an upper stage engine is smaller than a main engine, hence the chill down of the engine is significantly shorter (e.g. 15 min for the *Vinci* engine). But the chill down is performed under vacuum conditions, which means the altitude facility has a much longer operation time than the engine itself. The expander engine needs only one ignition system; it operates on flammable gas and can be activated several times. For regulation the expander engine has bypass valves to reduce the gas supply to the turbines (see Appendix G, Fig. G.1). One valve between the turbine and chamber valve on the hydrogen side mainly controls the combustion pressure. The second valve on the oxygen side mainly controls the mixture ratio.

The latter is completely opened during ignition and the other valve is almost closed. In this configuration the engine starts at a low operational point and the hydrogen pump receives almost all available power. During the chill down a certain

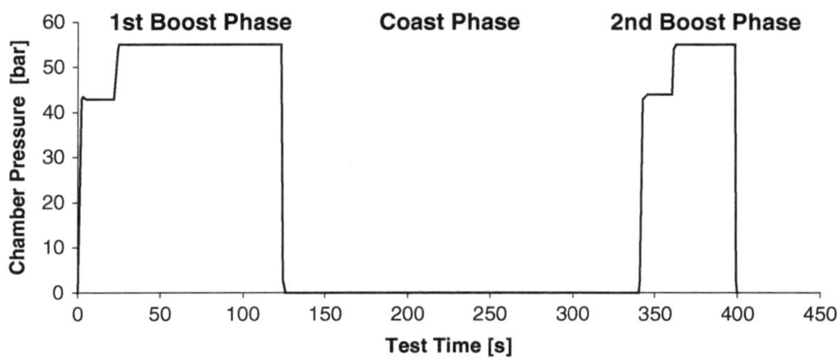

Fig. 4.12 Typical combustion pressure evolution vs time during a re-ignition test of an expander cycle engine test [9] (Photo: DLR)

amount of hydrogen has arrived in the heat exchanger (a *heat jacket* on the chamber wall), and this is heated up during ignition, expands and thereby makes the gas supply of the turbines and, respectively, the pumps [1]. The inlet pressure of the hydrogen pump during ignition is relatively high (6 bar). The pump, turbine and heat jacket are designed in a manner that the turbine provides enough power to run up the pump in this phase. For the shut-down of the engine both bypass valves are opened. By doing so the turbines lose their supply, the discharge of the fluids decreases, the chamber pressure and temperature drops down and finally the chamber valves are closed.

Chapter 5
Bench Systems

The test facility consists of all buildings, installations and systems within used for the test or in operational connection with the test (feed and purge lines, roof and court installations, for example). Examples of typical associated outbuildings are the control bunker, the fuel storage, the water tower and the steam generator of an altitude facility.

The engine test means an execution of an engineering process. Therefore the systems of the facility are classified according their relevance for test execution. There are systems for the process itself, for safety and for the building. The systems for the building (heating, air condition, lift etc.) are not part of the operational aspects of the facility. The safety aspects are only touched here where they have an impact on the process. The focus in this chapter is on the systems on the facility side which are directly needed for the test execution.

5.1 Principles for the Erection of a Test Facility

The main components of a test facility for a cryogenic rocket engine are the vessels for the fuel and for the oxidiser. These vessels are normally separated from each other and from the rocket engine by massive concrete walls (Fig. 5.1).

Remark 5.1 *During the erection of the facility* ***P5*** *at DLR in Lampoldshausen, in international cooperation the measures for a passage in a wall of 2 m thickness was misunderstood. In consequence the passage was at the wrong position and the prepared piping system had to be refabricated (Fig. 5.2).*

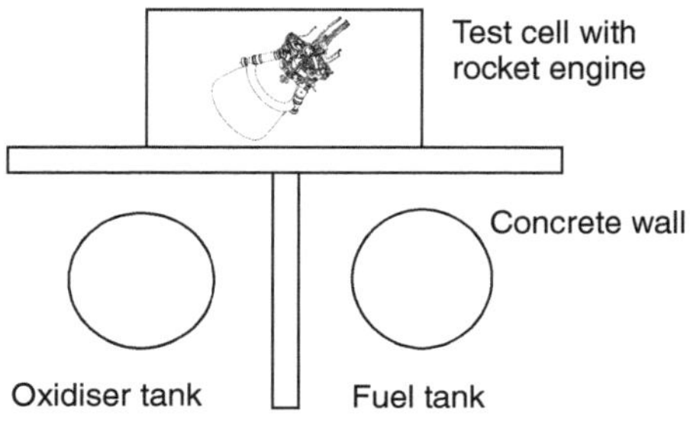

Fig. 5.1 Typical arrangement of the main components of a test facility (Photo: DLR)

W. Kitsche, *Operation of a Cryogenic Rocket Engine*, Springer Aerospace Technology,
DOI 10.1007/978-3-642-10565-4_5, © Springer-Verlag Berlin Heidelberg 2011

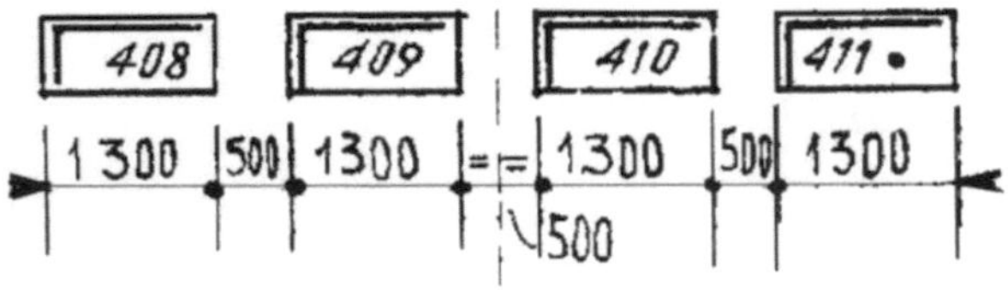

Fig. 5.2 Mistaken measurements. The French architects specified 500 mm between the two dots, the executing German company interpreted it as 2 × 500 mm because of the equality sign (Plan: SERETE)

5.2 Back Up Systems

The rocket engine as well as the test facility consists of several subsystems and components. In order to increase the reliability of the complete system important subsystems have a back up system.

Indispensable is a second computer system for control of the rocket engine. In case of a severe problem on the main computer during a hot run, and if the continuation of the test is not possible, a second computer has to be on stand-by to perform a proper abort of the test. An internal monitoring system of the computer (*watch dog*) watches all essential systems and functions of the computer, and switches the control automatically to the back up computer if one of the monitored functions is not provided any more (Table 5.1).

Indispensable as well is the electrical power. The total power supply of the test facility *P5* is about 800 kW (700 kW for the bench, 100 kW for the control building). Except for some very powerful consumers (e.g. water pump of 355 kW) all electrical consumers needed for the test are immune to a power failure. They are either connected to an *uninterrupted power supply* (e.g. the computer with 20 kW) where the power is taken from several battery chains or they are connected to a power stand-by unit driven by a *diesel* engine emergency backup generator. This engine and unit is started in case of power failure and runs up within 7 s.

Remark 5.2 *In former times computer systems had a much higher power consumption and heat release. Therefore the temperature in the computer room was also monitored, an air conditioning system (AC) having to maintain the correct temperature. The AC was able to transport a heat of 165 kW out of the control building, including its own electrical consumption of 45 kW. The AC also had an identical back up machine.*

Table 5.1 Back up power supply of the test facility *P5* at DLR in Lampoldshausen

Stand-by unit (Diesel engine)	NEA	398 KVA
Uninterrupted power supply	USV	3 × 60 KVA
For MCC and bench systems	230 V	3 × 62 Batteries
For engine valves	24 V	12 Batteries
For bench valves	60 V	2 × 13 Batteries

The principles regarding how to decide if a back up system is necessary or not are treated in Chap. 8. Another example for a backup function is the combination of sensor signals (see Sects. 4.4.3).

Remark 5.3 *The* ***Norsk Data*** *computer of the facility* ***P5*** *for the* ***Vulcain*** *engine merges all monitored systems and functions (e.g. power supply, internal data flow) into 12 collecting functions. After 4 years of test operation a failure occurred on this computer, a so-called* ***fugitive watchdog.*** *The failure message disappeared after some milliseconds. The message initiated correctly the immediate abort of the control of the facility. The back up computer checked the failure message in its cycle of 100 ms but found no failure message any more. Hence the back up did not take over control of the facility. It halted in its state like frozen. Thanks to god this did not occur in a hot run, because then the consequences would have been the exact opposite to a* ***frozen*** *test facility.*

5.3 Fuel and Oxidiser Supply

On many test facilities the rocket engine is tested in a manner of maximum analogy to the engine operation during flight *(static rocket system tests under rated and off design conditions)* [19]. In this case the run tanks of the test facility are dimensioned to provide fuel (and oxidiser) for a test time of 100–150% of the burning time during the flight. For testing cryogenic main engines that means the need of large run tanks (for example 600 m^3 and 200 m^3 for the *P5*). A fuel delivery by (road) tankers (Fig. 5.3) (up to 15 in the example above) with direct refilling of the run tanks would be a serious encumbrance to other facility activities. Ideal is the

Fig. 5.3 Trailer for liquid oxygen (Photo: DLR)

Fig. 5.4 Discharge of a LH_2 trailer (Photo: DLR)

reception of the (road) tankers at an intermediate storage (Fig. 5.4) and transfer of the cryogenic fluids from there into the run tanks. During the test periods the run tanks are regularly refilled, pressurised for the test, almost emptied in the hot run and again depressurised. Chill down and warm up means a cycle of thermal load applied to a tank which consumes its life time. Therefore complete discharging of the tanks is avoided in order not to warm up the inner jacket of the double-walled tanks and not to have more and more cycles of thermal load (Fig. 5.3).

5.3.1 Vacuum Insulation

All mentioned vessels and tubes are vacuum insulated, that means a double-walled tank or tube and the space between inner and outer tank (tube) is evacuated (e.g. at 10^{-4} mbar) (Fig. 5.5). This method is accepted as the best passive insulation and is typical of the test facility but for weight reasons rarely used on the launcher. The status of the vacuum sections has to be checked frequently and has to be corrected where necessary. A lack of insulation would deteriorate the chill down behaviour of the facility and the engine and could impede the reaching of the chill down criteria. The line between the run tank and the rocket engine has several (vacuum) sections. The evacuation of one section takes several hours. The inner tube normally has many sensors (for pressure, temperature, mass flow or vibration) but the vacuum section has no sensor. The best vacuum check is the on site inspection directly after the test. Experienced staff can identify by the condensed water (or even icing) on the line the status of the vacuum section. To recover the vacuum of a section is very time consuming, the section is connected to a turbo molecular pump; at first the connecting tube is evacuated and then, after hours of pumping, the vacuum section itself. The evacuation can last for days if the section is not at ambient temperature and even longer if humidity has ingressed. The ingress of humidity is absolutely to be

Fig. 5.5 Double walled LH_2 line with wire net protected compensators (Photo: DLR)

prevented and where appropriate the temperature of the section has to be increased with heating tapes.

The alternative to the evacuation is a filling of the section with carbon dioxide at ambient temperature. During the chill down of the inner tube the carbon dioxide transforms to dry ice and the pressure in the section decreases to some hundred mbar. This insulation method is not favourable on vertical lines. The gas would turn to ice, concentrate at the bottom of the section and create a bridge for heat conduction between inner and outer tube.

Remark 5.4 *Normally nitrogen is used to vent a vacuum section. A nitrogen gas bottle is connected to the section and it is vented slowly until the inside pressure has reached ambient pressure. By doing so a sudden increase is avoided and, much more important, no moisture can ingress the section. After many years of positive experience with this method a vacuum section of a partially filled hydrogen tank vas vented. The operator noticed with astonishment that the pressure in the section barely increased during venting. Doubtful, he asked for confirmation of the volume of the vacuum section. It became clear to the responsible engineer that the gaseous nitrogen in the section must have turned into ice and he gave the instruction to apply helium gas for venting. This instruction led to the next surprise for the operator. The helium remained gaseous and increased the pressure of the vacuum section rapidly. With a loud bang the pressure burst the disc which was installed on the section for safety reason (Figs. 5.6 and 5.7).*

5.3.2 Feed System

The feeding of fuel (or oxidiser) to the combustion chamber is mainly done by the turbo pump of the rocket engine itself. For proper performance of the pump the pressure at the pump inlet has to maintain a defined value above the requested *net positive suction head* (NPSH). For that purpose and to perform the chill down before the hot run, the test facility has a pressurisation system connected to the run tanks.

Fig. 5.6 Inner and outer flange of a double walled feed line (Photo: DLR)

Fig. 5.7 Vacuum pump stand with turbo molecular pump (*left*) and rotary vane pump (*right*) (Photo: DLR)

It is a powerful system with enough capacity to replace within the test duration the whole cryogenic fluid in the tanks by an adequate gas. On the facility *P5*, hydrogen gas is used to pressurise the hydrogen tank and nitrogen gas to pressurise the oxygen tank. These gases are stored at high pressure (e.g. 250 bar) and are reduced stepwise to low pressure (e.g. 3 bar) to float at last the run tank. Depending on the pressurisation mode and purpose heat introduction, evaporation of the liquids or condensation of the gas (see Appendix H) has to be considered. The pressurisation during the relative long chill down phase can create a more or less distinct layering of temperatures. The temperature deviation from the mean value is indeed only a few tenths of a degree but it can be decisive for the shut-down of the engine.

Remark 5.5 *For the first filling of the hydrogen feed line on the* ***P5*** *with liquid hydrogen, the test team pressurised the run tank and opened the valves towards a hydraulic dummy which simulated the rocket engine. It was assumed to be a simple calculation to find out what pressure in the tank would push the fluid through a standpipe to the inlet of the feed line. The test leader stopped the pressurisation when the calculated pressure was reached. But the temperature at the inlet of the feed line did not indicate any presence of a cryogenic fluid. Hence the pressure was raised in steps of 0.1 bar. For the sensors and the measurement chains it was the first time to measure the low temperature of the fluid, and it was uncertain that both worked without failure. After several steps of pressure increase the temperature dropped drastically and the typical 20 K of liquid hydrogen was indicated. Afterwards we figured out that a small amount of liquid at the inlet instantly evaporates and a remarkable counter pressure is created which pushes the liquid back into the tank. This evaporation creates a pressure oscillation in the feed line during the first phase of the chill down. The amplitude of this oscillation is of the order of the mean pressure value in the feed line. The phenomenon is treated in the literature as the* ***surge*** *effect. It can also occur on a launcher and has to be treated there with much more attention because the flight line is far more fragile than the bench line (Fig. 5.8).*

First attempts to compute the pressure oscillation in the feed line were made by the author in 1994 (Fig. 5.9). The feed system was modelled in one dimension of

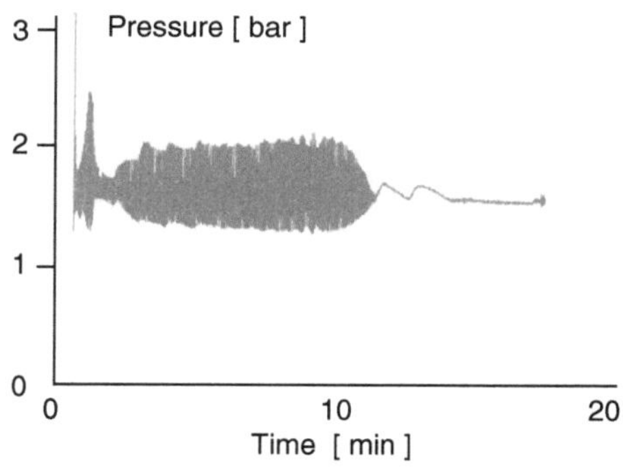

Fig. 5.8 Pressure oscillation in the hydrogen feed line during the first phase of the chill down (Photo: DLR)

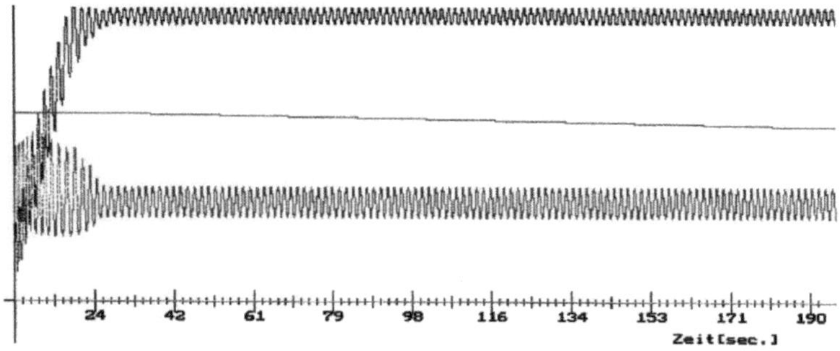

Fig. 5.9 Early computation of the first phase of a chill down pressure (*bottom*) and filling level (*top*) of a hydrogen feed line vs time (ScreenShot: DLR)

space and in time (unsteady). The physics of the model is described by a system of ordinary differential equations (ODE) and was computed by means of the *Kutta-Runge* algorithm. The results were in sufficient agreement with a good part of the measurements and with the principle behaviour of the system.

The transient behaviour has essential importance for all liquid space propulsion systems. Therefore the ESA supported a program to develop a software library for computation of various processes in fluid systems of spacecrafts. [13]

Remark 5.6 *The test facility P8 for the development of high pressure combustion chambers at DLR in Lampoldshausen reaches at the interface bench/specimen a maximum pressure of 320 bar. Because the specimens have no turbo pumps, the enormous pressure has to be created by pressurisation of the run tanks. The tanks have an internal shield which prevents an injection of the gas into the cryogenic fluid. The first attempts at pressurisation failed because the velocity of the gas was too high and its energy dissipated. The temperature of the fluid increased to an undesirably high level but the pressure could not be increased sufficiently. The test leader developed a* ***parabolic*** *pressurisation in which the pressure in the tank was increased within 1 min along a parabolic profile from 1 to 500 bar.*

5.4 Measurement, Control and Command (MCC) Systems

The need for precise test execution, monitoring of important parameters and exact measurement is as old as rocket engine testing. The test engineers of the past always applied the available state of the art in technique for this purpose. Nowadays these tasks are done by a *measurement, control and command (MCC)* system (Fig. 5.10). It normally consists of a system of computers, storage devices and a periphery as the interface to the facility and engine and for operator work at terminals. Meanwhile the MCC is not only applied for the hot run but also for test preparation and for the reset of the facility after test.

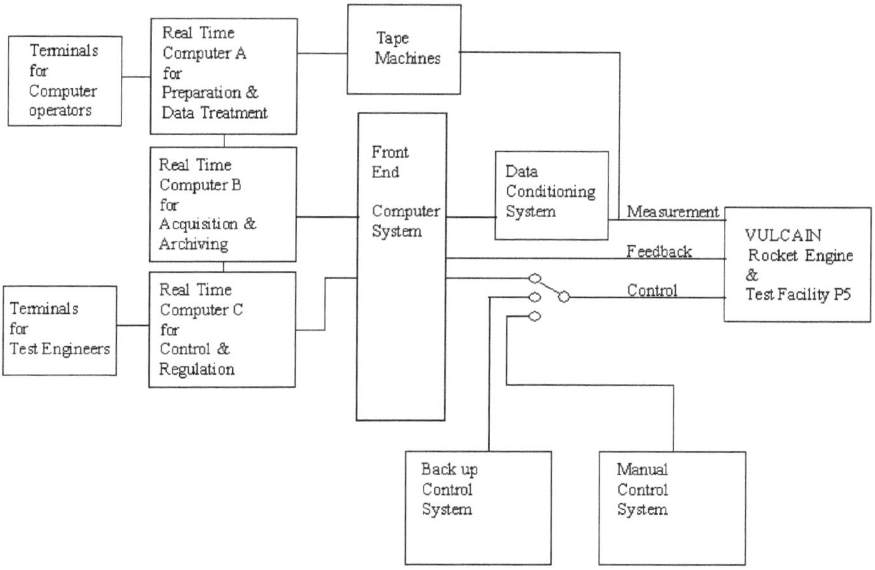

Fig. 5.10 Original MCC system of the facility *P5* (Photo: DLR)

5.4.1 Historical Review

In the first test facilities for rocket engines the piping went through the control room. The test engineers adjusted the pressure reducers and regulation valves manually and read off the values from pressure gauges. The bench and the specimen were observed through a slot in a concrete wall or through a periscope.

Remark 5.7 *The pioneers of rocket engine testing only had a small fraction of today's control systems. Mostly they applied the principle of starting a pre-adjusted engine and running it in one operational point without any variation up to the shut-down. The second generation of test engineers corrected a valve position from a safe distance by means of cords. To the amazement of the scholar of the old school, a perfect mixture ratio levelled out during the hot run.*

Due to the development of electronics and sensors in the 1960s and 1970s more and more non-electrical measurements were transformed and displayed on analogue gauges. The control room of a test facility of those days was dominated by this kind of instrument (Fig. 5.11).

Each engineer of the test team watched about four gauges and informed the test leader in cases of reaching a critical value; then the test leader decided for test abortion or continuation. The sight to the engine was still through the bulletproof glass in the slots of the bunker wall. By and by the monitors entered the control rooms of the test facilities. At first they were used to display pictures from cameras. Parallel with the development of the computer, the display of data on a screen entered the control

Fig. 5.11 Control room of the test facility *P4* for the *Viking* engine (1970s) (Photo: DLR)

systems. The arrays of desks as in the NASA control centre in Houston became the standard control technique for space application (Figs. 5.12 and 5.13).

In the 1980s the available electronics and display components were consequently applied. Remote controlled cameras transmitted high quality coloured pictures to several monitors in the control room. All valves and pressure reducers necessary for test guidance were remote controlled from a switch board of several metres width. Special monitors with *light pens* provided all the required measurement values in a displayed piping synoptic. The computer allowed a higher degree of automation. In the 1970s the computer only controlled the hot run but in the 1980s the test

Fig. 5.12 NASA control centre in Houston during the *Apollo* program (1960s) (Website: NASA)

Fig. 5.13 Control room of the facility *P5* for the *Vulcain* engine DLR, Lampoldshausen (1980s) (Photo: DLR)

preparation was also performed with computer support. The test engineer no longer actuated a remote controlled valve; instead he entered a command via the keyboard of his terminal and started a complex sequence of valve activations. In the 1980s each valve and sensor still had its own cabling between the test bench and the control bunker.

In the 1990s the large switch boards disappeared from the control rooms and the desks were replaced by normal tables with standard monitors and keyboards. These terminals were connected to large computers still with the structure of the old main frame computer. The extensive cabling between control building and test bench was drastically reduced by the application of intelligent systems on the facility. These systems processed the measured signals and transferred digital data in collective lines to the MCC in the control bunker.

The computers of the current MCC systems have the structure of a workstation and the acquisition of data is via high frequency signals converted to digital format, no longer analogue on magnetic tapes (Fig. 5.14).

5.4.2 Control Programs

In the 1970s the control based on fix electronic hardware components was replaced by programmable logic controllers (PLC). In the beginning only digital inputs and outputs were available, which were read and set in a cycle. The number of inputs and

Fig. 5.14 Control room of a test facility in the 1990s (Photo: DLR)

outputs was extendable with modules and thereby the number of channels was high enough even for complex facilities. The cycling time (e.g. 2.5 ms) was also short enough to control even fast processes. The processing of analogue signals (e.g. limit switches) had to be realised by additional, special components. Later generations of these control units became more powerful and were also able to perform regulation processes.

Remark 5.8 *The DLR applied the PLC (a Procontic by BBC) on small test facilities until 1997. Besides the Procontic, more and more* ***Siemens*** *computer were installed (S3, S5 and S7) (Fig. 5.15).*

The typical computer for control of a test facility in the 1980s was a *main frame computer* or a set of *main frame computers*. This kind of computer runs programs of different categories. The basic is an operating system, and on top of this we have the utilities to operate the computer (*software and hardware*). For the control (including regulation and monitoring) of the rocket engine and test facility, another category of programs was applied, the sequences.

A sequence contains a number of statements which are sequentially processed. A good part of the statements is standard in any program language (mathematical operations, loops, subroutine calls etc.); further on there are statements for the dialog with the periphery of the control computer. Their purpose is to switch a valve, activate a system, read a sensor signal, release a reference value etc. The program language in the 1980s was a derivative from FORTRAN; in 2008 it was rather like PASCAL. Figure 5.16 shows a loop from a purging sequence which opens/closes valves with defined waiting times in between.

An example of a simple use of a sequence is the testing of valves. Selected valves are opened/closed by the sequence in a defined timing. In parallel the position feed back from the valves is recorded. The evaluation means a study of opening/closing duration of the valves.

On a test day, between 70 and 80 sequences are applied, partially in parallel, partially in series. The sequences used make up a structured system (Fig. 5.17); they

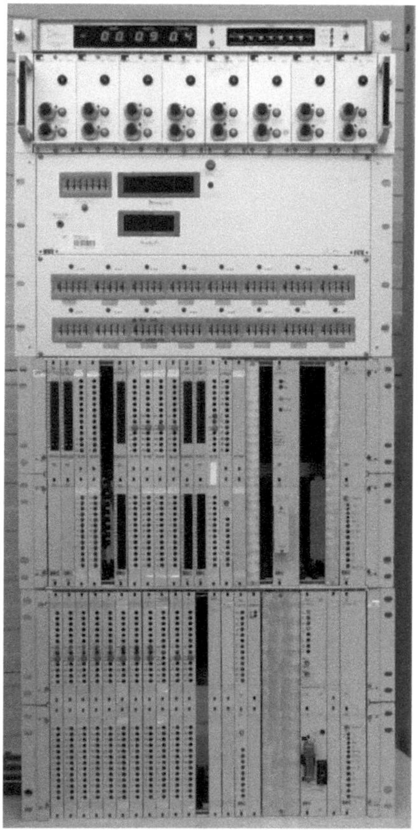

Fig. 5.15 Programmable logic controller, BBC (Photo: DLR)

have different priorities and a specific task. Sequences can start and stop each other, there are conditioned and unconditioned activations of sequences and they can enter a stand-by state. The name of the sequence contains significant information about its priority, the controlled area and the task of the sequence. The sequence named

```
COMPUTE COUNTER = 0
DO
   EXIT ON COUNTER > 9
   SWITCHOFF AVY250,0,ACTION=ALARM
   WAIT DELAY = 40S
   SWITCHOFF AVY126,0,ACTION=ALARM
   WAIT DELAY = 20S
   SWITCHON AVY250,0,ACTION=ALARM
   WAIT DELAY = 40S
   COMPUTE COUNTER = COUNTER + 1
   MESSAGE 3,1,1001,COUNTER
   SWITCHON AVY126,0,ACTION=ALARM
ENDDO
```

```
for I := 1 to 10 do
        AVY250(FALSE);
        SeqWaitTime(T#40s);
        AVY126(FALSE);
        SeqWaitTime(T#20s);
        AVY250(TRUE);
        SeqWaitTime(T#40s);
        LogMsgCon1 := CONCAT("-I- ",(I));
        LogMsgTxt := CONCAT(LogMsgCon1,".
Bedrueckung LH2 mit GHe!");
        ILogMsg(0,GetCallers(),LogMsgTxt);
        AVY126(TRUE);
end_for;
```

Fig. 5.16 (**a**) Part of a sequence 1988. (**b**) Part of a sequence 2008 (Sequence: DLR)

Phase (chronological order)	Main-Sequenz	Sub-Sequence				
Automatic Sequence (H0-5')	NETSTDQA	NETSTPRP (NETSTPR1)	NZLDEBCD			
			NEPRPRLS			
			NZPARDIS			
		NETSTSTA (NETSTST1)	NESTARLS			
			NEBUBO2M			
			NHDISPH2			
			NODISPOX			
			NEPCSA5S (Gimballing)	UECHDEND	NHPURTPH	NHPURLH2
						NHSMOTPH
						NHPURVAH
					NOPURTPO	NOPURRIE
						NOSMOTPO
						NOPURVAO
			NESCPOFG			
			NEABOSTA	NZCHDCRT		
			UEWATOFF	NEABOSTA	NZCHDCRT	
			UEENGSDN	NZBCC1		
				NZBCC2		
				NZBCC3		
				NZBCC4		
				UEENGRLO	UEBENRLO	
				UEAKRBUT		
				NETSTEND	NHDISPH2	
					NODISPOX	
					NHPURTPH	NHPURLH2
						NHSMOTPH
						NHPURVAH
					NOPURTPO	NOPURRIE
						NOSMOTPO
						NOPURVAO
					NZNAOFF	
		NEENGMON	NEGIMBAL			
			NEENGREC			
Hot run (H0)		NEENGSTA (NEENGST1)	[NZBCC8]			
			[NZBCC9]			
			[NZRHOLOH]			
			NEH02PFX			
			NEH02PFY			
			NEEVPPO			
			NEEVPPH			
			NEEVPPHB			
			NEENGREG	NZRHOLOH		
				NZREGMAN		
Shut-down (H2)			UEENGSDN	NZBCC1		
				NZBCC2		
				NZBCC4		
				UEENGRLO	UEBENRLO	
				UEAKRBUT		
				NETSTEND	NHDISPH2	
					NODISPOX	

Fig. 5.17 List of sequences for a *Vulcain*-Test (without chill down and reset phase) (Photo: DLR)

NHPURLH2, for example, has a **N**ormal priority, controls the **H**ydrogen area and has to **PUR**ge the **LH2** area.

Often a sequence is executed in dialog to the test team. The sequence, for example, offers the responsible operator a prolongation of the chill down phase of a segment. Due to this the operator can synchronise his process to other processes on the facility. The most drastic external influence comes from the *red button* of the test leader desk. This button releases immediately a pre-selected shut-down of the rocket engine and a transient of the facility into the safe state.

The *hot run sequence* has a central function within all test sequences. While this sequence is executed no intervention from the test team is possible (except via the red button). For control of the hot run the precision of control signals is of particular importance. In this phase precision cannot be taken for granted because, especially here, the signals are close-packed. For the release of a control command in the 1980s a maximum delay of 10 ms was allowed. Depending on the type of command the computer took up to 6 ms to execute it. If several commands were released at the same time (e.g. during start up or shut-down) the delay of the last command increased to values longer than 10 ms. But for many systems (and for valves) this was still acceptable because the reaction time of these systems itself was generally more than one order of magnitude higher (see Remarks 3.2 and 3.3).

5.4.3 Engine Control

The standard control element on a rocket engine is the pneumatic open/close valve (e.g. before the combustion chamber). The pneumatic actuator is driven by an inert gas (e.g. helium) and the open/close chambers are pressurised/depressurised via electromagnetic valves (pilot valves). The electrical activation of the pilot valves comes originally from the MCC system of the facility (see Appendix M). For hydraulic valves (e.g. thrust vector control) the same principle is applied but they have the advantage that arbitrary valve positions between minimum and maximum are adjustable. Normally each valve has a feed back of its position which is read as a digital input (for regulation valves as an analogous input) by the MCC system. The pyrotechnical elements to ignite the engine are themselves ignited by initiators (a pill of powder) which are ignited electrically. These electrical igniter circuits are also activated by the MCC system authorised by key switches. By the coordinated actions of all control elements and the commands of the MCC all functions of the rocket engine are controlled (start up, switch off, regulation, thrust vector adjustment, ground-to-flight switch over etc.). The commands for the control elements are released periodically with a precision of $^{-0}_{+10}ms$. The MCC system of the facility for the *Vulcain* tests had three linked computers each with several processing units. The periphery of the MCC system has interfaces for the engine and facility control, for data acquisition and recording and the terminals to operate the total facility (MCC system, test bench and engine) (Table 5.2).

Table 5.2 Input and output channels (bench and engine) of the MCC system of the facility *P5*

Input channels	Number
Analog	608
Impulse counter	10
Digital	2048
HF data (20 kHz)	96
Output channels	
Analog	10
Digital	512

5.4.4 Data Acquisition

The analogous signals from the sensors are transmitted via measurement chains (see Appendix L) to a *signal conditioning system* and forwarded to the A/D converters in the periphery of the computer. The signal conditioning system is a number of amplifier racks where the signal of the sensor is adapted to the input of the computer. The MCC system controls and, respectively, sets the amplification and offset of the sensor signal in the conditioning system. Internally the computer stores a raw value of the measurement and computes according to a polynomial the physical value of the measurement. The power supply of the sensor (see Appendix K) is also located within the signal conditioning system. The long way of the cabling between signal conditioning (in the control bunker) and the sensor (on the facility or engine) induces a drop of voltage. Therefore, beside the original measurement signal, the voltage at the sensor is also measured via a sense line. Altogether a total of six wires are needed for one sensor (measurement, sense and supply circuit). Most of the sensors measure pressures or temperatures on the facility or rocket engine; other measured physical parameters are vibrations, forces, deformations and shaft speeds (Fig. 5.18).

Fig. 5.18 Signal conditioning system (amplifier racks) (Photo: DLR)

5.4.5 Measurement Validation

The acquired raw data are very seldom directly used for regulation or for the monitoring of parameters. The data for such a purpose go through a validation process. A typical validation of data for regulation of the *Vulcain engine* is the following process:

1. Computation of the mean value of the last five acquired values
2. Exclusion of all values that differ more than 20% from the mean value
3. Computation of the mean value of the remaining values

4. Selection of the first mean value in cases of exclusion of all values (see 2 above), and selection of the second mean value if no more than four values were excluded

Remark 5.9 *Among the aerospace systems since the 1970s, electronic regulation units have displaced more and more mechanic/hydraulic regulation units. Until then the non-electrical parameters were boosted hydraulically or mechanically and then directly used for control purpose. If electronic units are used the non-electric parameters are converted by sensors into an electrical signal, the regulation unit reads the signal, processes it and sends another electrical signal to electric/mechanic actuators. The standard signal interface in the 1970s was the* ***V24 interface****. In this regulation the engine engineers had to face, for the first time, the problem of* ***non-physical measurement values****. A regulation which completely relied on the correctness of the used signals caused unwanted shut offs and regulation failures. The reason was a disturbance of the signal, often only for very short times (****peaks****). It duped a violation of a limit concerning a critical engine parameter to the regulation unit. This kind of error can be reduced by improvement of the hardware or (more favourably) by improvement of the programming of the regulation algorithm (measurement validation).*

5.5 Detection Systems

While a test facility is operational its vessels are permanently filled with fuel, oxidiser and complementary fluids. To avoid a warm up the vessels for the cryogenic fluids are not completely emptied in a test. The complementary fluids are partially stored at high pressure (e.g. 300 bar). The risk of a leak in a vessel or in the piping cannot be excluded completely. Therefore the facility is equipped with gas detectors around the tanks and in all rooms where pipes are routed. In the case of a gas leakage or displacement of air the detectors linked to a monitoring system will automatically set off an alarm. In addition to these gas detectors the system is equipped with differential heat sensors. The close surroundings of the rocket engine is monitored with two gas analyser systems, each working on another physical principle (e.g. oxygen-reduction-reaction, see Appendix N).

Inside the rocket engine further gas concentration measurements are taken. They reveal a possible internal leak during the chill down phase. The gas samples are taken with *Teflon* hoses which burn off after ignition.

Large test facilities are evacuated before authorisation to open the main valves is given. Because from that moment on no personnel are present on the facility, all systems have to be not only remotely controlled but also remotely monitored. Besides monitoring by detectors, monitoring by cameras is also very important. Events which cannot be sensed by the detectors might still be visible on the video monitors in the control room.

The risk of the presence of hydrogen or oxygen due to an internal leak inside the lines is minimised during the non-active phases of the bench by a safety pressurisation of the lines with inert gas.

The correct function of this pressurisation and the tank pressure are monitored in non-active phases. A failure initiates an alarm and releases the organised communication of the responsible operators.

5.6 Test Cell Systems

The test cell houses the rocket engine in its test position. During the hot run it is wide open, that is the large lid of the exhaust guiding system is removed, the wide lateral gates are opened and the armour plated doors to the other rooms of the test bench are closed. The device to move the large lid and the wide gates are test cell systems, the schedule of the test day directly depending on their correct function. The thrust frame which bears the engine forces is another system of the test cell; it is a heavy, massive steel construction which is anchored in the concrete structure of the bench.

Between the engine and the inlet of the exhaust guiding system, three pilot burners are installed (Fig. 5.19). Their purpose is to ignite hydrogen which enters the test cell from the moment the chamber valve is opened. Without these burners the hydrogen would ignite with the engine ignition and the detonating gas would burn

Fig. 5.19 Test cell with running pilot burners and cooling water supply, in between the torus with angled nozzles for creation of an ejector jet(Photo: DLR)

around the engine where cables and insulation material could be affected. These pilot ignition systems are typical for cryogenic engines: a gas flame (for the *Vulcain*) or a spark providing device (e.g. for the *Space Shuttle Main Engine*) is used for that purpose. Another test cell system inhibits the climb up of the burnt gas, a suction system running on an ejector jet at the inlet of the exhaust guiding duct and a curtain like jet around the engine creating a downward flow into the wide duct.

5.7 Exhaust System

In its nominal operational point the *Vulcain* converts more than 3 GW chemical power into thermal power. All structures and systems in the vicinity of the exhaust jet have to be adequately protected from this thermal load. The launch table for the rocket has to bear this load for less than 10 s and it is sufficient to cover the concerned area with a water layer. On the test facility the exhaust jet has to be guided away from the facility and the area concerned is loaded with the thermal power of the exhaust jet for several minutes. Test facilities erected on a hillside with enough depth below the engine have the advantage that the jet can blow onto the surface at the bottom (Fig. 5.20).

Fig. 5.20 Test facility without exhaust guiding system, test stand 1A at air force research laboratory in Edwards, USA (Website: US Air Force)

A guiding system is necessary if the exhaust jet has to be led away from the facility. A large duct guides and deflects the jet. Such a system is made of ordinary steel and definitely needs a cooling system (Figs. 5.21–5.23).

The hillside position of a facility is also an advantage for the design of the cooling water supply of an exhaust guiding system. The system is fed by water towers which are erected at a higher geodetic point than the facility. Due to the hydrostatic pressure the installation of water pumps of high power can be at least partially reduced.

Remark 5.10 *The cooling water flow for the* ***Vulcain*** ($P_{thermal} = 3,3\,GW$) *was 2,000 L/s, 46% of the amount evaporated. The higher power of the* ***Vulcain 2*** *required an increase to 2,750 L/s and a significant higher fraction evaporated.*

The exhaust guiding system of the facility ***P5*** *including its cooling water supply was the biggest permanent challenge in the operation of this facility. After each test, an inspection and normally small repairs are necessary.*

At the beginning of the test periods the water distribution was optimised. Enough water has to be present everywhere; on the other hand, too much water causes a partial blocking of the duct.

During one of the first ***Vulcain 2*** *tests an upper segment of the system fused and, during the first long duration tests of the* ***Vulcain****, soak water was rejected from the upper segment into the test cell. The water hindered the sight to the engine and*

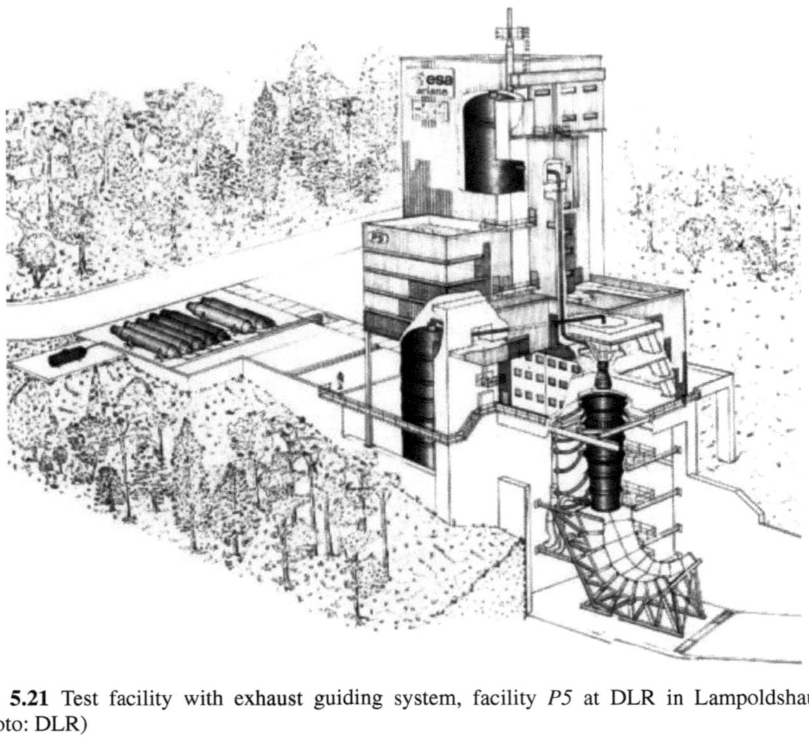

Fig. 5.21 Test facility with exhaust guiding system, facility *P5* at DLR in Lampoldshausen (Photo: DLR)

Fig. 5.22 Typical temperature distribution in an exhaust guiding system (Screen Shot: DLR)

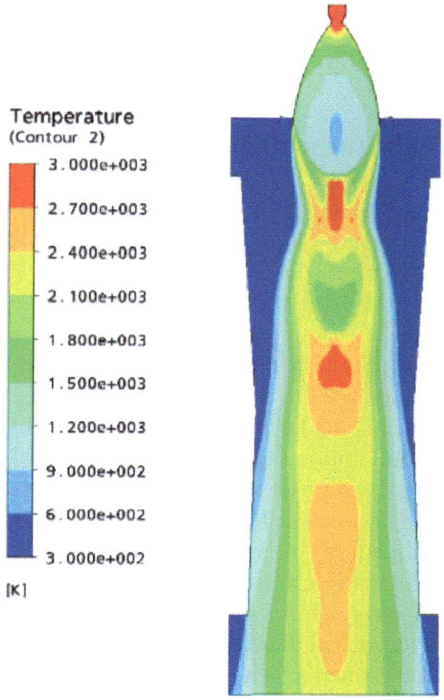

another severe problem emerged, the soak water also reached the nozzle; it caused a partial cooling of the nozzle, the water was evaporated and the nozzle returned again to its expected temperature. This could happen several times in one hot run and the nozzle designer had to take each of these events as a hot run cycle of the nozzle. Due to that fact the life time of the nozzle was drastically reduced.

The exhaust guiding system has to withstand the highest load when the thrust vector control of the engine is applied during a hot run. The oblique exhaust jet

Fig. 5.23 Flow check of the exhaust cooling system (Photo: DLR)

of the gimballed engine hits the inner surface of the guide tube. In order to avoid melting of the steel it can only be pointed at the same spot for a short time.

5.8 Altitude Simulation

Upper stage engines are started and operated during a flight phase at high altitude. The conditions of these altitudes, in particular the low pressure, are simulated in test facilities with vacuum chambers. The test cell of such a facility is designed to be evacuated and the pressure can be lowered to some mbar. But the low pressure also has to be maintained during engine operation and therefore a suction system is required to suck off the complete engine exhaust. The vacuum chamber and the suction system upgrade the test facility to an *altitude facility* in which upper stage engines can be adequately validated for their mission. The vacuum in the test cell before ignition can be established slowly by means of pumps, but for engine operation a more powerful system is necessary. The exhaust jet of normal upper stage engines (e.g. as for the *Ariane 5*) are sucked off by ejectors. They work on the same principle as the water jet pump (see Appendix I). The jet of the ejector is created by combustion of a fuel/oxidiser mixture at high pressure and addition of water (steam generator). Except for the water injection the steam generator is in principle a rocket combustion chamber working at considerable pressure and temperature. The created steam jet is used to operate the ejector and, respectively, to suck off the engine exhaust. To obtain a sufficiently low pressure the facility has multiple stages (Table 5.3). The steam is back liquefied in a giant condenser by injection of cold water and collected in an underground water basin. The diffuser below the engine and the following duct of the exhaust and ejector jet is a highly interesting subject of gas dynamics but still the design can only be realised successful in an accurate development with fundamental experience. The reason for the complexity of this flow are the effects of phase and heat transition, the different *Mach* number regions, turbulence, boundary layer effects and the still ongoing chemical reaction. Hence the prediction of the flow and the calculation of the overall performance of the altitude facility is an ambitious subject. The altitude facility including its steam generators (Fig. 5.24 does not show the five steam generators of the facility) is an enormous increase of the complexity of the overall test facility. In principle a second facility

Table 5.3 Performance of the steam generators for the altitude facility *P4.1* at DLR in Lampoldshausen

Steam generator	Ejector	Mass flow	Steam pressure	Operational time
1	1	55 kg/s	20 bar	15 min
2	1	55 kg/s	20 bar	15 min
3	2	58 kg/s	20 bar	15 min
4	2	58 kg/s	20 bar	15 min
5	Chill down ejector	10 kg/s	20 bar	25 min

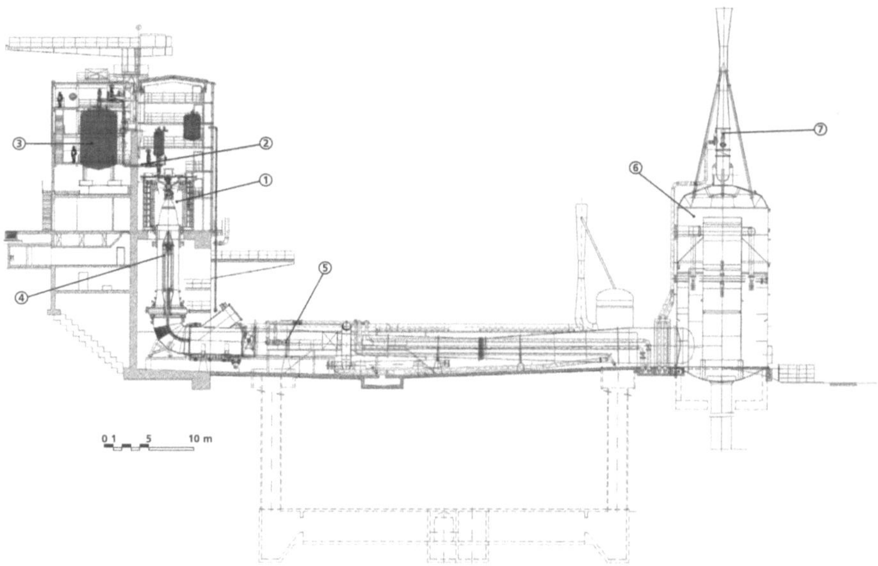

Fig. 5.24 Altitude facility *P4.1* at DLR in Lampoldshausen for the development of the *Vinci* engine under altitude condition (Photo: DLR)

is operated which is to be adjusted precisely at its interface to the rocket engine in the test cell.

On facilities for a small thruster (e.g. 10 N) the vacuum test chamber can be kept at low pressure (1–30 mbar) by means of big vessels (e.g. 500 m^3). The vessels are evacuated by Roots pumps and rotary vane pumps. Pump operation during a firing test maintains the vacuum which means that the test duration is not limited by the facility. Such a facility can also be used for short duration tests (e.g. 20 s) of small engines of some hundred Newton of force.

Another alternative is the operation of ejectors with steam created before the test and stored in vessels.

Chapter 6
Simulation of Flight Conditions

A basic requirement of the test facility is to apply as far as possible flight conditions to the tested rocket engine [10]. It is mandatory that all interfaces between the bench and the engine are adjusted to the same values as on the launcher. Furthermore the evolution of interface parameters versus the time is regulated according the evolution during flight.

6.1 Pump Inlet Pressure Profile

Before the start of the launcher a sufficient pressure at the inlet of the rocket engines turbo pumps has to be established. During lift off this inlet pressure (Fig. 6.1) suddenly jumps and increases further during flight, due to the increase of acceleration. The pressure loss in the feed line slightly inhibits this pressure increase but it becomes really significant when the launcher has powerful boosters such as, e.g. the *Ariane 5*.

Up to ignition of the engine the same procedure as on the launcher is followed on the test facility, but in order to simulate the conditions during flight the pressurisation of the run tank has to be adapted. To obtain the requested pressure profile a powerful pressurisation system (Fig. 6.2), an effective depressurisation and a fast throttle in the feed line is necessary.

The pressurisation system for the tank on the launcher maintains a certain pressure p_{TL} while the tank runs empty during flight. At the inlet of the rocket engine the pressure p_I is

$$p_I = p_{\mathrm{TL}} + \Delta p_{\mathrm{hyd}} - \Delta p_{\mathrm{dyd}} \tag{6.1}$$

where Δp_{hyd} is the hydrostatic pressure due to the difference of height between tank and engine and Δp_{dyd} is the pressure loss in the feed line. On the test facility *P5* the feed line has the same design as on the launcher, especially the same length. In order to obtain the same inlet pressure on the bench, here the tank pressure p_{TB} has to be higher compared to the launcher:

$$p_{TB} - p_{\mathrm{TL}} = \rho a l. \tag{6.2}$$

W. Kitsche, *Operation of a Cryogenic Rocket Engine*, Springer Aerospace Technology,
DOI 10.1007/978-3-642-10565-4_6,

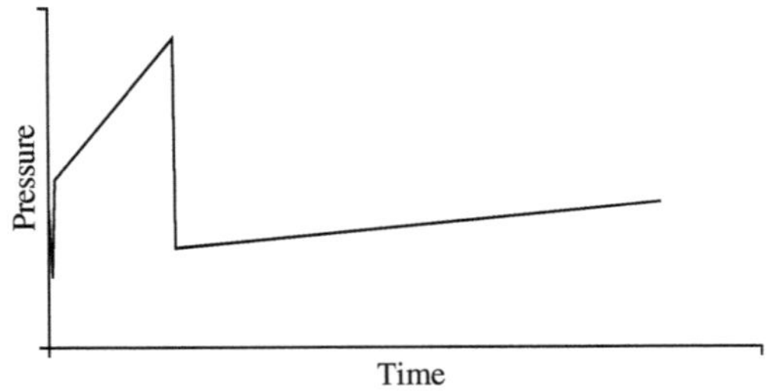

Fig. 6.1 Principle pressure profile at the LOX pump inlet during flight (Photo: DLR)

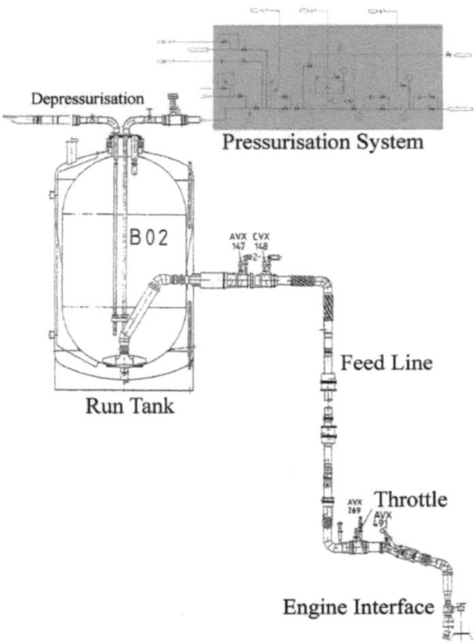

Fig. 6.2 Liquid oxygen supply of the *Vulcain* engine (Photo: DLR)

This difference depends on the increasing acceleration a during flight, the density ρ of liquid oxygen and the length l of the feed line, its value increasing up to 14 bar. This pressure concerns the whole run tank and for that reason it is very strong and heavy and so it got its name *battle ship tank*.

On a test facility running on liquid hydrogen it is preferable to use pneumatic actuators (e.g. N_2 driven) for the valves. For the fast throttle in the main feed line the performance of a pneumatic actuator was not sufficient, so a hydraulic actuator with high opening/closing speed was chosen. By the normal pressure relief valves the depressurisation was not fast enough so an additional relief valve of 250 mm diameter was installed. Figure 6.3 shows the pressure at the inlet of the pump of the rocket engine. Clearly visible is the pressure drop after ignition, during start up and the sudden pressure increase when the throttle fully opens. The real effect on the launcher (acceleration) means fewer disturbances than the simulation (fast opening of the throttle) and hence we can consider that the test on the bench sufficiently reveals the behaviour during this phase.

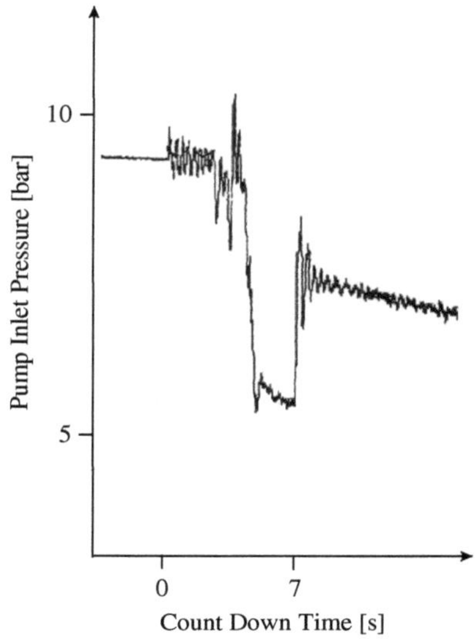

Fig. 6.3 First phase of the engine hot run (Photo: DLR)

The inverse event occurs 2 min later, during shut-down of the *Ariane 5* boosters. The inlet pressure of the main engine drops drastically. Figure 6.4 shows the simulation of this pressure drop on the test facility.

No cavitation was observed, the water hammer and oscillation were never critical for the engine, so neither the simulated ignition nor shut-down of the boosters had a critical impact on the operational behaviour of the rocket engine or its components.

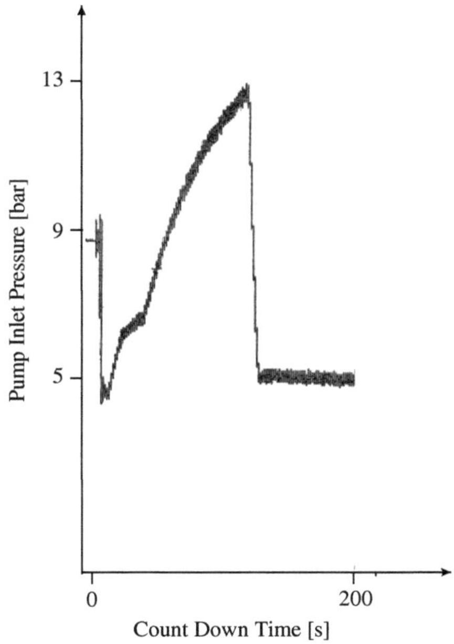

Fig. 6.4 Pressure drop at the inlet of the oxygen turbo pump during simulated booster shut-down (Photo: DLR)

6.2 *Pogo* Oscillation

Since large cryogenic launchers are applied, the rocket scientists have to face the phenomenon of *pogo* oscillation [15]. The liquid oxygen in a long feed line of a rocket engine tends to show this oscillation when evaporation creates gas bubbles in the line. The effect is well known and a simple damper (Fig. 6.5) can reduce the oscillation.

However, the pogo effect does not always occur and to prove the effectiveness of the damper an oscillation is generated with an additional hydraulic device. The first tests on the facility *P5* were performed with a strong feed line which was double walled and vacuum insulated. This feed line for liquid oxygen had no damper; later on the feed line was replaced by a flight line, insulated with hard foam, with the original design for flight, including a system to damp *pogo* oscillation.

Oscillation with strong amplitudes can be expected directly after engine shut-down (Fig. 6.6).

The flight line with the damper has in principle the same behaviour but of course the oscillation dies down a bit faster.

After it was proved that the oscillation after shut-down, meaning after a sudden pressure *increase*, was not critical, the oscillation after a sudden pressure *drop* was investigated (Fig. 6.7). A pressure drop always means a change in direction of the vapour pressure and this might induce the risk of evaporation of the cryogenic fluid in the feed line. Meanwhile a system (hydraulic piston) to generate oscillation was connected to the feed line and it was activated after the simulated booster shut-down.

The oscillation was of course stronger when the generator was applied, but it died down directly after the generator was stopped and no critical excitation occurred while the generator was running.

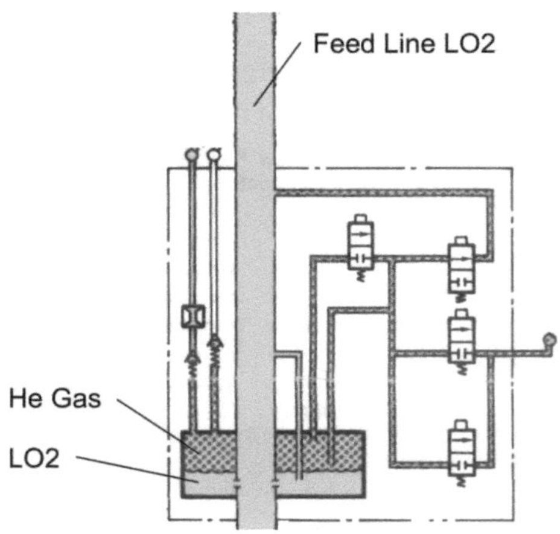

Fig. 6.5 Damper for *pogo* oscillation (Plan: SNECMA)

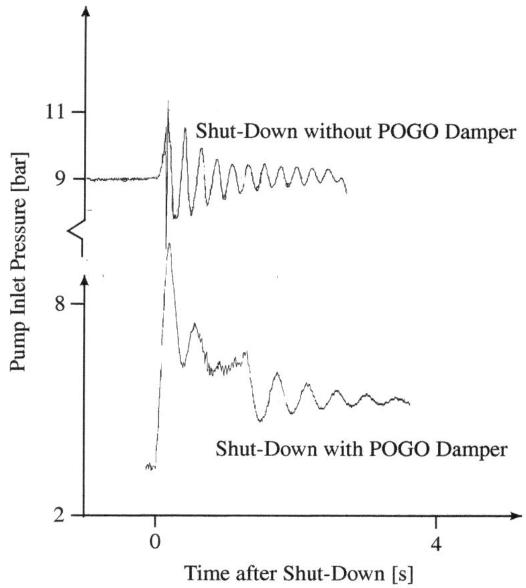

Fig. 6.6 Oscillation in the feed line after engine shut-down (Photo: DLR)

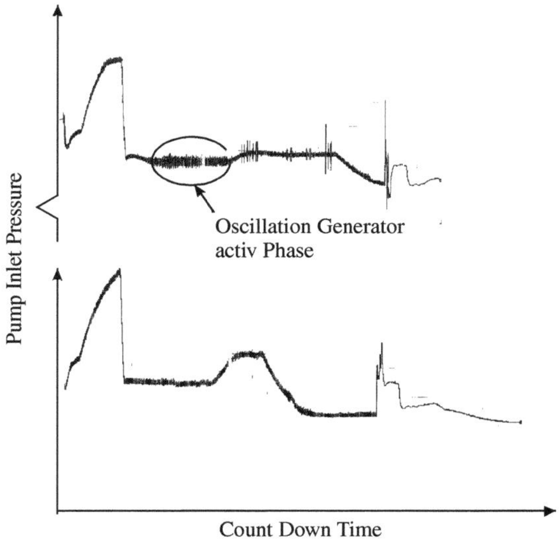

Fig. 6.7 Hot run with oscillation generator (*upper curve*) and without generator (*lower curve*) (Photo: DLR)

6.3 Acoustic Load

Another condition due to the booster operation during the first 2 min of flight is the acoustic load caused by the booster jet. The main engine in the middle has to suffer this enormous noise and it was questionable whether all components including tiny lines, cables and plugs would bear this load without failure.

Fig. 6.8 Acoustic panels around the *Vulcain* (Photo: DLR)

The required sound level was higher than the performance of the acoustic chambers available in Europe, hence the idea was to use the *Vulcain* itself as the acoustic source and to trap the sound in a casing (Fig. 6.8) around the engine, made of strong steel plates.

The strong steel panels reflect the sound back to the engine, where it is reflected again and permanently more noise is produced. In these tests the influence of the acoustic load on a *running rocket engine* was studied and it was considered to be a much better simulation of the flight conditions than it could have been in an acoustic chamber, in which a hot run is not possible.

No failure of any engine component or negative impact to its function was observed.

Remark 6.1 *The temperature (Table 6.1) closely surrounding the running rocket engine (not below the nozzle) is quite moderate. Considering that many components of the engine conduct the cryogenic it is not surprising that their surface temperature is still* $-20\,°C$*. But the noise around the engine is so high that a person in that area would suffer injury. Figure 6.9 shows sound measurement at the level and circumferential position of the oxygen turbo pump at 3 m distance from the centre of the rocket engine. The consideration is, however, theoretical, because the presence of anyone in this area during a hot run is already forbidden for safety reasons.*

Table 6.1 Temperature measurement during a *Vulcain* test

Distance from the nozzle	Height above the nozzle rim	Temperature
350 mm	800 mm	380 K
1600 mm	375 mm	480 K
2500 mm	375 mm	420 K

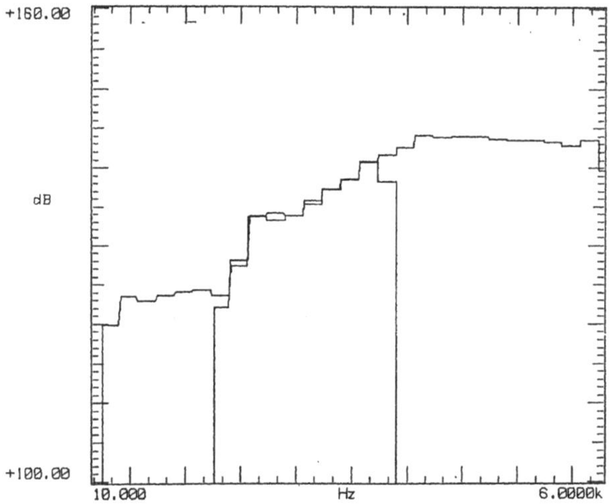

Fig. 6.9 Sound measurement during a *Vulcain* test (Photo: DLR)

6.4 Nozzle Load

The test facility *P5* has further systems to simulate flight conditions and was temporarily modified to perform tests with conditions which are difficult to simulate. The most sophisticated case was a *load simulation device* (LSD) to scrutinise the load on the engine nozzle at high altitude. It is described in detail in [3].

During flight the ambient pressure around the nozzle permanently decreases; hence the load on the nozzle increases and the simulation of this load became the subject of a test series. Another load during flight is caused by the transition from subsonic to supersonic flow. The launcher reaches supersonic velocity while it is still in the atmosphere. Even before reaching *Mach 1* the complex flow field around the launcher including the engine has regions with sonic shocks. Furthermore, the flow is not steady and, accordingly, not the shock waves. The effect of moving, appearing and disappearing shock waves causes a buffeting to the structure and is named after this the *buffeting effect*. The simulation of this dynamic effect was beyond the possibilities of a test facility. The only thing which was done to gain certain knowledge of this effect on the engine was to introduce a force to the exit of the nozzle by a hydraulic piston. The force was constant and only of the order of the forces expected during flight.

Normally, a rocket engine for the first stage is tested at sea level conditions only. Altitude facilities with vacuum test cells are typical for upper stage engines. Hence there was no facility available to test the *Vulcain 2* under altitude conditions when a failure of this engine occurred during its first flight.

The investigation after the failure of the *Ariane 5* flight number 157 identified a weakness of the nozzle extension to be the origin of the problem. It was known before that the load during flight, especially at high altitude, is higher than at sea level, but the load was underestimated and therefore the nozzle had to be replaced by a stronger one. To qualify the new design, a special test campaign was performed which was concentrated on the behaviour of the new nozzle. Therefore, the facility *P5* was equipped with a casing around the nozzle and a suction system (see Appendix I) to reduce the pressure around the nozzle. The main part of the suction system was an ejector driven by nitrogen. The tightness and fast depressurisation were a particular challenge when this device was used. Also in these tests the conditions on the launcher were to be copied as far as possible. Therefore it was requested that the nozzle is not disturbed in its thermal behaviour due to the load simulation device. Hence it was necessary to consider a cooling system for the casing around the nozzle from the beginning of the design onwards (Fig. 6.10).

To reach a load equivalent to the load during flight, a pressure of 200 mbar in the casing was requested. The request was not only to reach this point, but also to perform a specified profile of pressure reduction vs time. Therefore the suction system was planned with two regulation valves of different diameter. To create a requested sudden recovery of ambient pressure at shut-down, the casing was equipped with a burst discs, bursting after ignition of small detonators.

The casing of the device was connected by a lattice ring (Fig. 6.11) to the steel structure of the bench. The top and the bottom of the casing had a seal to close the cavity for maintaining low pressure inside.

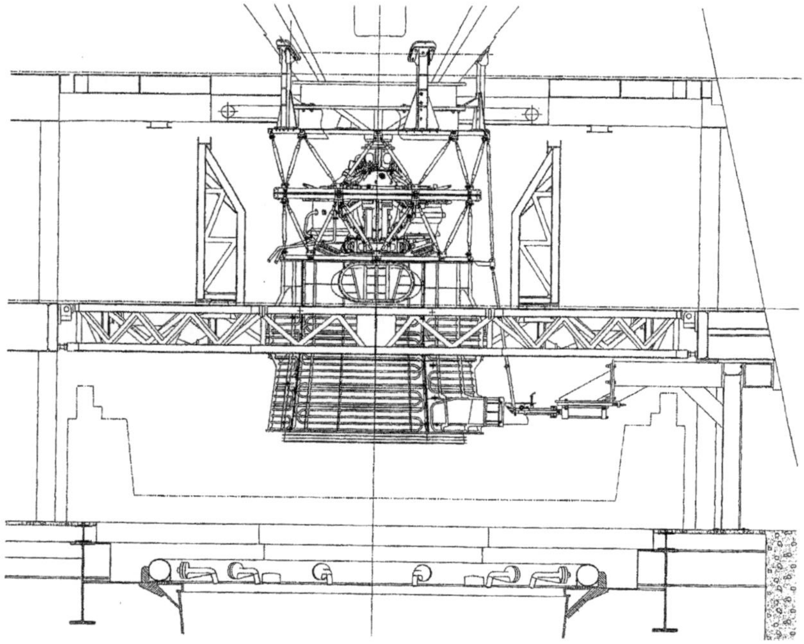

Fig. 6.10 Test cell with *load simulation device*, LSD (Photo: DLR)

Fig. 6.11 Vacuum casing of the *Load Simulation Device*, LSD with water cooling during hot firing (Photo: DLR)

The seals had to be flexible to allow the normal movement of the nozzle during a hot run. It had to be resistant to fire and strong enough to bear the static and dynamic forces during test. As material, a texture of fibreglass was chosen. A particular problem was the fixing of the seal which needed stepwise optimisation. A remaining leak rate across the seal was considered and measured in the first reception phase of the LSD. Moreover, the leak raised another problem: the leak might have had a fraction of hydrogen in the air, and therefore attempts were made to adapt the bench measurement systems to measure a hydrogen leak at low pressure.

At the end of the reception phase, the characteristics of the suction system including the influence of the leak flow was known (Table 6.2).

In the final reception of the system, the ejector was driven with 39.6 bar and created a vacuum of 9 mbar before the suction line was opened. This value increased

Table 6.2 Performance of the LSD suction system

Suction pressure in the vacuum casing	200 mbar
Leak ratio into the casing	3 kg/s
Driving mass flow of the suction system	14 kg/s GN_2
Driving pressure of the suction system	40 bar
Suction pressure at closed suction line	14 mbar

to 115 mbar when the valves opened to adjust 300 mbar in the casing of the LSD. After 58 s, the leak ratio increased but the system maintained the value of 300 mbar. After the LSD had its *go for test*, it was used during hot runs with a lowest pressure of 216 mbar around the nozzle extension.

After this test and the following inspections of the test material, the improved nozzle was validated for flight and the *Ariane 5 ECA* could return to mission.

Chapter 7
Weather Conditions

Large test facilities accommodate most of their systems in a proper building, sometimes even air conditioned. Nevertheless many systems are still installed in the open air. The weather conditions have an important influence on the performance and function of many bench systems.

7.1 Ambient Temperature

The overall lower temperature limit in the specification of a test facility is typically around 0°C. This does not mean that the bench is closed below this temperature, but that the performance is reduced below this temperature and special attention has to be paid for certain systems and, if a test has to be conducted, special measures have to be taken.

The most obvious system which is inhibited in its function at temperatures below zero is the exhaust cooling system. The water supply ends in tiny tubes and nozzles all over the inner surface of the exhaust guide tube. If the water starts to freeze the supply is interrupted and local and temporary overheating of the guide tube must be expected.

If the test facility has an open test cell, all test cell systems have to cope with the weather conditions. A typical test cell system around a cryogenic rocket engine is a burner system to ignite hydrogen escaping prior to engine ignition. Burners operating on propane will have problems around 0°C because the vapour pressure becomes low. Countermeasures can be taken, such as temperature regulation of the propane tank.

Remark 7.1 *In the second week of January 1995 the client urged the **P5 Vulcain** team to conduct an important test for the **Ariane** program. The ambient temperature in the morning was −7°C, (Fig. 7.1) the colleagues from the winter service had carried out their job very well, and the test team prepared the subsystems and performed pre-tests. Despite a good temperature adjustment in the propane tank the burner could not be set into operation. The piping system was too cold and caused re-liquidation of the propane at the end of the line.*

W. Kitsche, *Operation of a Cryogenic Rocket Engine*, Springer Aerospace Technology,

DOI 10.1007/978-3-642-10565-4_7,

Fig. 7.1 Test facility during winter (Photo: DLR)

Low temperatures diminish the tightness of seals and packing sets and wherever ice is created the risk of blocking (e.g. valve pistons, venting holes) has to be considered.

The function of evaporators is also strongly dependent on the ambient temperature. If the evaporator is covered with snow or ice the heat transfer falls dramatically and we also have to consider that even above 0°C ambient the evaporation of cryogenic fluids can turn humidity into ice on the surface of the system.

7.2 Rain and Humidity

Rain normally causes no problem on a test facility. If an outside system has a rain-sensitive component, it has extra shelter (e.g. a plate above a pressure gauge on an external vessel). Only in failure cases, if rain water enters the building and effects, for example, an electrical system, can a problem occur.

Remark 7.2 *On the rocket engine the presence of water is particularly undesirable. Even in cases of burning material on the engine the test leader avoids the use of the fire fighting system because water damage to the electronics, plugs and sensors is much more serious than burnt insulation or cables.*

Similar to rain, a high humidity in air can affect electrical systems. A flare stack with its ignition system on the very top of the facility is completely exposed to any weather conditions and can experience ignition problems due to rain and humidity.

7.3 Thunder and Lightning

If a complex system like a launcher on the launch pad or a test facility is struck by lightning, the system can suffer severe damage. Lighting conductors (Fig. 7.2) are installed but the risk of an anomaly cannot be completely eliminated.

When not being used a test facility is more robust against a disturbance due to lightning than during test conduction. Therefore the rule on a test facility is to suspend the test preparation and even postpone the test if thunder and lightning are around the test area.

Remark 7.3 *In former times all the facilities had the measurement devices in the control building. They were connected by long cables to the sensors on the test specimen. The measurement cables must be screened and the screen has to be earthed. This configuration means a weak point and the kind of earthing has an important influence on the sensitivity of the measurement circuits against lightning.*

Fig. 7.2 Girder masts as lightning conductors on the *Ariane 5* launch pad (Photo: Author)

7.4 Sight and Fog

Fog and darkness is relevant to the safety and security management of the test centre during test operation. The hazard area has to be monitored via cameras by the safety engineers. If visibility is no longer sufficient due to fog or darkness, surveillance is no longer possible. In this case external persons unaware of the risk might intrude and any abnormal behaviour of facility systems would not be noticed.

7.5 Wind

Even strong winds are normally no problem for the test facility itself. But wind moves the exhaust of the rocket engine. Therefore the wind speed and direction is very important in figuring out where the exhaust might come down. The cryogenic rocket engine produces pure water and this exhaust is much less critical than the exhaust of rocket engines running on storable propellants with toxic fractions.

The wind also influences the propagation of the sound or rather noise of the rocket engine. A dry winter day and a leafless forest facilitates sound propagation and a layer of cloud at a certain altitude can reflect the sound back to the ground.

7.6 Atmospheric Sound Propagation

The influences of atmospheric winter/summer conditions are also confirmed by *infrasound measurement*. Infrasound measurement enables the detection of sound events caused by meteorite entry, re-entry of spacecrafts, volcanic eruptions, explosions in the atmosphere, as well as rocket launches and rocket testing.

An infrasound measurement array of the *International Monitoring System* (IMS), the station IS26 in Bavaria, Germany, at a distance of 320 km, recorded the sound emission of several *Vulcain 2* tests on the facility *P5*. Figure 7.3 shows signals between 3 and 9 Hz caused by a test in December 2001. The signal window is wider than the real test time which indicates that the sound is propagated not only along the surface but also along a stratospheric path [12]. On the other hand no signals were detected from any of the tests conducted during the summer.

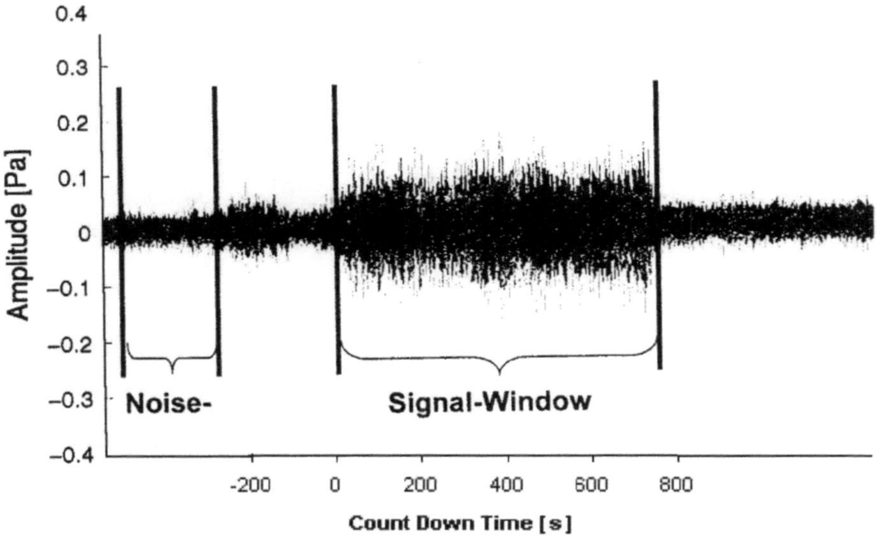

Fig. 7.3 Infrasound measurement of a *Vulcain 2* test (Photo: IMS/DLR)

Chapter 8
Test Procedures

The application of prescribed procedures is typical of the work on a test facility [6]. This guarantees the traceability and reproducibility, it creates detailed documentation and therefore enables optimisation of the procedures and the work. This again accounts for high work efficiency and for a high level of safety and quality.

8.1 Test Readiness Procedure

Before the start of a test campaign the test requirer reviews the configuration and documentation of the scheduled specimen and he compares the *limitations and constraints* of the rocket engine with the objectives *(test plan, test request)* of the scheduled tests. The operator of the test facility reviews the technical configuration of the facility and compares the performance limits of the facility to the test request. The result of these reviews is a technical conclusion which is presented in a test readiness meeting before the campaign. In principle the test readiness of the rocket engine and the test facility is confirmed in this test readiness meeting but, on the basis of the latest technical results, the test readiness has to be confirmed again shortly before each test comes up for execution. For that purpose the test requirer checks the results of the previous test and compares the compatibility of the foreseen operational point and operational mode vs the assigned limits *(e.g. a forecast vs redline margins).* The test readiness procedure ends with the test readiness meeting on the test day where all criteria (see Sect. 4.2) relevant for the test readiness are reviewed. The meteorological conditions (wind, ambient temperature, probability of storms and lightning) are also part of these criteria. Then the test has to be executed within the same hardware and software configuration as the earlier dry run. That imposes a repetition of the dry run if one or more criteria are not reached and a modification of the configuration is necessary. Normally that means a postponement of the test of at least 1 day.

Nevertheless, after the test readiness is confirmed a test abort is still possible if an anomaly occurs which makes the continuation impossible or if the continuation induces an unacceptable risk.

W. Kitsche, *Operation of a Cryogenic Rocket Engine*, Springer Aerospace Technology, DOI 10.1007/978-3-642-10565-4_8, © Springer-Verlag Berlin Heidelberg 2011

Remark 8.1 *On 12.12.2002 the* ***Vulcain 2*** *test* ***M209-03*** *was planned on the facility* ***P5****. The evening before the* ***Ariane 5 ECA****, flight 157 had failed, the first flight of a* ***Vulcain 2****. Hence the morning after the first question of the test leader to the test requirer was about the test green light, and he received a* ***go for test****. So the test team continued the test preparation and, at the end of the test readiness meeting, the test leader and the test requirer put their signatures to the minutes. The chill down of the bench lines started at 12:49 h, and at 13:01 h the test leader received a phone call in the control room in which he was asked by the test requirer to abort the final test preparation.*

The next test was performed on 16.07.2003 after a modification phase of 7 months (see Sect. 6.4).

8.2 Non-conformance Procedure

The configuration of the facility and its adjustment for the test are defined in the relevant documents. A deviation from this definition has to be classified as a non-conformance. The non-conformance can be a technical failure (defect of a system or a component) or a human error (e.g. wrong adjustment). Every person in the test team is obliged to report any non-conformance immediately. The report is made on a coloured, eye catching form on which the essential information about the observed anomaly is fixed. This *non-conformance report* (NCR) is transmitted within hours to the responsible quality inspector. Between the tests an *NCR review board* regularly meets to treat the anomalies (*Commission de Traitement des Anomalies, CTA)*. The minimum composition of this board is the test leader, test requirer and a quality agent; it can be enlarged arbitrarily by specialists. The task of the board is failure assessment, to define measures to resolve the failures and to follow up the corrective measures. Before conclusion of the test readiness meeting the board has to classify by signature all NCR into one of three categories: the failure is resolved, the failure is acceptable for test or the failure is not acceptable for test. In the last case no clearance for test can be given. In a normal test period (30 tests per year) about 1,000 NCRs are opened on a facility like *P5*. Because each NCR has to be treated several times, and in spite of the board meeting at least twice between tests, the agenda of the meeting can easily have 40 NCRs. Therefore proper preparation of the meeting and a skilled treatment of the NCRs are very important for the test schedule. In addition to the use of forms, computer application became more and more important, particularly for statistical exploitation and the documentation of failures (experience memory).

The occurrence of NCRs during the life time of a large test facility is shown in Fig. 8.1. Typical is the increasing frequency in the beginning, up to a high maximum and then a subsequent asymptotic decrease to a minimum. In general the frequency increases slowly on an old facility but this increase can be avoided by targeted replacements or upgrades of subsystems (see Sect. 7.4).

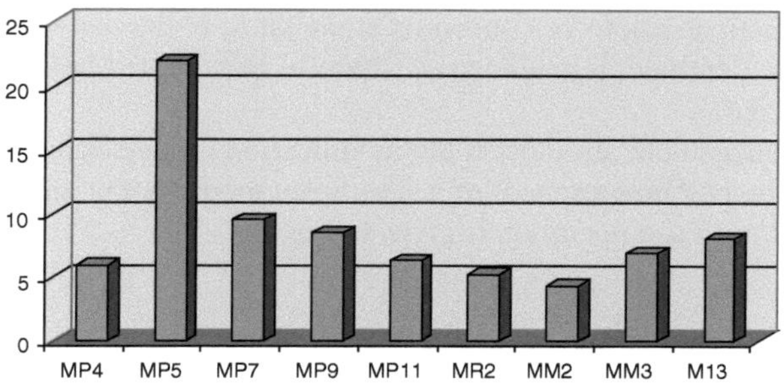

Fig. 8.1 NCRs per test during several campaigns on a test facility for a rocket engine (Statistic: DLR)

Remark 8.2 *Naturally several or many colleagues are involved in a procedure on the test facility. Therefore it is important to reach a good level of knowledge and acceptance of a procedure among all the persons involved. Especially important is a good introduction of the procedure which should be arranged by the leader of the team himself and should be given to the complete team.*

*On the **P5** a senior colleague approached the test leader because of an NCR which concerned his working field. Well experienced in erection and commissioning but with little knowledge of the meaning of the non-conformance procedure he complained with the red NCR form sheet in his hand: "We've arranged the whole facility for you and didn't we do a good job? And now you show us the **red card**?"*

8.3 Risk Management During Operation

The risk management of the test facility in its different phases of existence is treated in Chap. 8. During the operation phase additional, repetitive risk analyses are performed. The purpose is to treat risks which arise from modifications of hardware or the test process.

Basically a risk analysis is performed for every new subsystem on the facility. A risk analysis before each test campaign is standard and, in addition to it, an analysis before every test.

The reference situation of the test facility (including test procedure) for the analysis is the configuration during the prior test. All modifications of hardware (facility and rocket engine) and software (test procedure) and all identified non-conformance are systematically studied. Accordingly, the identified risks counteractions are defined and their proceedings are followed up. The target is to reduce the risks and, respectively, to eliminate them. The remaining risks are evaluated related to their acceptability. Thereby the type of risk (personal injury, hardware damage, loss of test objectives) and the probability of occurrence are considered. The success of the counteractions has to be confirmed at the latest in the test readiness meeting (Sect. 8.1). All identified, remaining risks have to be accepted before the *test green light* can be given.

The risk management also covers the modification processes on the facility. The link between the risk management of a significant modification and permanent risk management during test operation is given in Fig. 8.2.

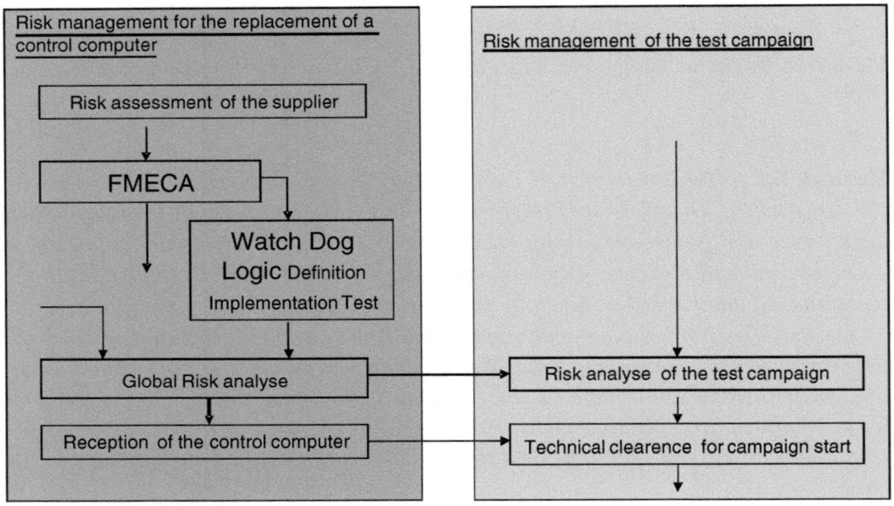

Fig. 8.2 Risk management for the replacement of a control computer (Screen Shot : DLR)

8.4 Configuration Management

Concerning the configuration, the test facility is divided into systems for the test process, for the building and for safety purposes. For the systems of the test process a defined configuration management is applied. The configuration is documented in corresponding files and plans. If the test request changes it is a priori not certain that, within the limits of adjustment of the test facility, the requested test configuration can be reached. In particular, when the type of specimen is changed, modification of the facility can be necessary. Modification of the facility is performed and documented according a prescribed procedure. Any rebuilding, extension or new installation of hardware or software means a modification of the facility. The modification procedure consists of four main steps:

1. *Description of the modification request*
   The identified need of the modification is justified and a proposal for remedy is made. In principle, everyone in the test team can release this request.

2. *Treatment by the modification board*
   The board consists of the test requirer, test/project leader, technical specialists and assistants. The board assesses the request, decides about the go ahead and follows up the realisation of the modification.
3. *Project Phase*
   The responsible specialists define the details of the modification (specifications, tender documents), commercial clerks invite offers, offers and plans are verified and the project leader establishes a planning of the realisation.
4. *Award and execution of the modification*
   The selection of the contractor is made according a defined procedure and finally justified in a *choice report*. According to the specification and the offer (including the technical part), the contract is awarded and the execution is followed up by specialists and quality assurance. The execution phase is normally accompanied by special *key points* (e.g. reviews, approval of drawings) and progress meetings. The acceptance of large modifications can be divided into up to four phases.

The delivery of long lead items or large and complex systems have to be considered in long term planning. The expected delivery date is normally announced and confirmed by the supplier in progress meetings. Space systems are often unique and sophisticated. Therefore the procurement can have delays (Fig. 8.3). Progress meetings and permanently updated planning for relevant deliveries and activities are useful tools during commissioning as well as during facility modification.

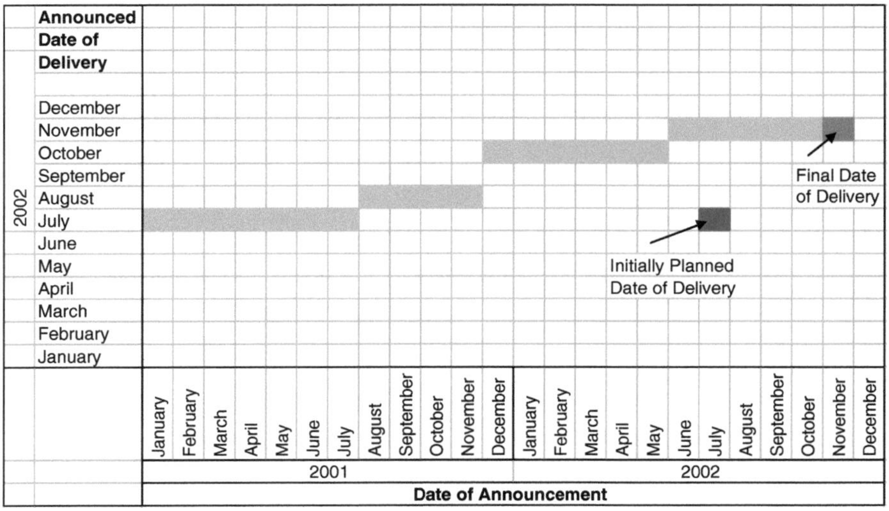

Fig. 8.3 Example of the course of an expected delivery date during the procurement phase (Screen Shot : DLR)

Remark 8.3 *In the progress meetings, not only are the delivery dates of the items confirmed but also the level of progress of all items in the workshop. During commissioning of the* **P5** *the client was pleased with the progress of a compressor system having a level of 15, 30 and then 60%. But in the next meeting suddenly a status of 15% was declared. The site manager of the supplier explained to the surprised audience that the 60% only concerned the status of the drawings and that the real production had not yet begun.*

Chapter 9
Safety

The operation of a test facility for a rocket engine implies several risks. The large amount of liquid hydrogen and oxygen, high service pressures and the enormous density of energy when the rocket engine is running requires systematic risk management (Sect. 8.3), the integration of adequate safety systems and their prescribed application within the facility and test operation.[1]

The facility operation is analysed related to the expected risks, risks are classified and counteractions for risk reduction/elimination are performed. The risk management covers all phases of existence of a facility (strategic planning, design, erection, modification, reception, operation, shut-down and removal).

In cooperation, the operator, the constructor and the responsible authorities establish a safety concept during the strategic planning phase. An essential term in this phase is the *TNT equivalent*. The facility operator studies the worst failure case on the facility (in our case the explosion of fuel) and calculates for this event the equivalent mass of the explosive agent TNT which would have the same impact. According to this level of risk, the constructor designs the planned buildings and safety radii, hazard areas and explosion proof areas are usually defined.

In parallel to the design, the analysis (RAMS, FMECA, SPF e.g.) concerning the safety status of the facility is performed and the necessary safety equipment is defined. Accordingly this equipment is integrated during erection. During reception and operation of the facility the safety aspects are considered within the test procedures.

The concept of the facility has to ensure a safe configuration in the rest position as well as after complete shut-down. The removal of the facility must not be linked to unacceptable risk.

In this book we focus on the safety aspects during operation and normal rest (night, weekend).

---

[1] Within this book we do not treat the common aspects of job security and safety.

W. Kitsche, *Operation of a Cryogenic Rocket Engine*, Springer Aerospace Technology,
DOI 10.1007/978-3-642-10565-4_9,

Where cryogenic fluids are applied one rule always needs to be considered:

Cryogenic fluids must never be locked without control

Of course a cryotank can be closed during operation (e.g. for self pressurisation) but the pressure increase going along with the locking of such a fluid always has to be considered. For that reason every lockable line segment and every tank is equipped with its own safety equipment (safety valve, burst disc). Additionally the concept of the facility automatically ensures a safe rest position in case of loss of power supply or loss of pneumatic control without an activation of safety equipment of the individual component.

The fundamental safety principle of testing a large rocket engine is the *test performance on a completely evacuated test bench*. Further on the hazard area (Table 9.1) around the bench has also to be evacuated, prohibited and guarded. Personnel are only allowed in a safe building (bunker).

As well as the systems for the test process, the components of the safety system are remote controlled from the safe bunker. According to the phase (rest or test) the safety system is set into the adequate mode. In the test phase, fire detection in the test cell must not release an alarm or a fire fighting system. For the test and safety systems the principle of *homogeneous redundancy* (back up equal to main system) is applied. A request of *diverse redundancy* (back up different to main system) would also apply to the software and processors (dissimilarity) as normal in aerospace. On the test facility only a back up system with equal or less capability than the main system is installed. This philosophy is justified because the facility is remote controlled and an emergency shut-down cannot be excluded which is acceptable on a test facility. It is requested for all systems used in the test process that the failure of a single component (*single-point failure*) [20] does not induce a malfunction of the overall system. On the other hand, the facility is not safeguarded against a double failure which means if main and back up systems fail a subsequent error is accepted.

The support of a safety officer is requested for every test. He operates a fire fighting system (Fig. 9.3) and a gas and fire detection system; subsequently the fire brigade of the test centre is on stand-by (Figs. 9.1 and 9.2).

Table 9.1 Safety restrictions for the facility *P5*

Time related to engine ignition (H_0) respectively to shut-down (H_2)	Activity on the bench	Test bench status/Safety zone around the bench
$H_0 - 8\,h$ to $H_0 - 3h$	Test preparation	Access only for the test team
$H_0 - 3\,h$ to $H_0 - 2h$	Chill down of the feed lines	Test bench evacuated
$H_0 - 2h$ to $H_0 - 20\,min$	Chill down of the engine	250 m safety radius
$H_0 - 20\,min$ to $H_2 + 30\,min$	Hot run (H_0 to H_2)	400 m safety radius
$H_2 + 30\,min$ to $H_2 + 2h$	Purging/venting of engine and feed lines	Test bench evacuated

Fig. 9.1 Station of the fire brigade (Photo: DLR)

The team leaves the bench before the tank valves are opened. The clearance for bench access after a test is given when feed lines and the engine have been cleaned with inert gas and when all monitoring (gas and fire detectors, cameras) indicates a safe situation on the bench. First access after test is the start of a safety inspection of the complete facility. In the pre-, post- and inter-test phase, personnel are present on the bench. Therefore several safety systems are very important in this phases (pressurisation of feed lines, tank pressure monitoring, O_2-, H_2-, fire detection, pressurisation of the building). These systems are also active in the rest phase

Fig. 9.2 Fire practice of the fire brigade (Photo: DLR)

Fig. 9.3 Water gun for fire fighting (Photo: DLR)

(night, weekend); a malfunction in this phase releases a prepared safety procedure including a call to the responsible bench operators. In this phase, on the one hand the risk for the personnel is reduced or eliminated; on the other hand, reaction time after an alarm is longer.

Chapter 10
Documentation

The complexity of a space program means that the documentation is also vast and complex. A large part of the documentation is not only for internal use but has to be distributed to other partners in the space program. Therefore the agencies have defined a structure and a numbering system for official documents. The document number indicates first of all the program (e.g. A5 for *Ariane 5*) and the type of document (e.g. SM *specification de management*). This definition is laid down in a technical flow chart; it has management character and therefore has the number A5-SM-1000000-P-20. The management of documents is meanwhile computerised, the applied software working not necessarily with the official numeration but with an extra number or with the internal number of the company. Therefore the internal number is also used in the heading of the documents (Fig. 10.1).

The documents related to the test facility can be divided into those for the erection and acceptance (facility documents, such as build drawings, reception files, etc.) and those for the operation (procedures, numerical codes, test requests, etc.) of the facility. As this book focuses on the operational aspects of the test facility, this chapter deals mainly with the documents used in a test period.

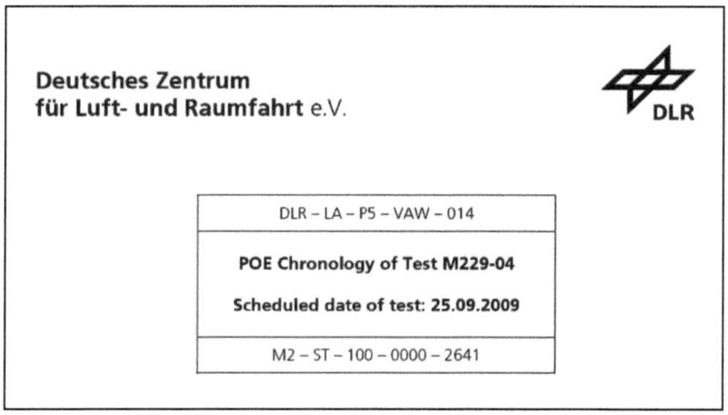
Deutsches Zentrum
für Luft- und Raumfahrt e.V.

DLR

DLR – LA – P5 – VAW – 014

POE Chronology of Test M229-04

Scheduled date of test: 25.09.2009

M2 – ST – 100 – 0000 – 2641

Fig. 10.1 Heading of a test document in the *Vulcain 2* program on the facility *P5* at DLR-Lampoldshausen (Documents: DLR)

W. Kitsche, *Operation of a Cryogenic Rocket Engine*, Springer Aerospace Technology, DOI 10.1007/978-3-642-10565-4_10, © Springer-Verlag Berlin Heidelberg 2011

10.1 Facility Documents

The test facility is documented in the same manner as any other technical facility. Every subsystem is documented by piping diagrams or cabling plans, user and maintenance manuals, data sheets, specifications, etc. The documents are available on paper and more and more they are also stored electronically.

10.2 Test Documents

During a test period of a *Vulcain* engine on the *P5*, 12 official documents are regularly exchanged between the test requirer and the facility operator (Table 10.1).

10.2.1 Test Request

The central document during a test period is the test request *(Demande d'essai)*. It defines the tested hardware and the test objectives, contains a forecast of the operational point, gives all directions for the preparation, execution (including emergency shut-down) and post processing of a hot run. It also includes the request for the measurement points and monitoring. The author is the test requirer; in the *Ariane* program it is the prime contractor for the rocket engine. The creation of this document requires detailed and special experience concerning the rocket engine and the test facility. In practice the test requirer and the facility operator are in permanent communication in order to gain the test definition vs the limitations of the facility. The result of this dialog is a *Complementary Test Request* which completes the test definition.

Table 10.1 Documents in a test period

Document	Content	Author
Test request	Specification of all planned tests in a campaign	Test requirer
Test request	Specification of the next test	Test requirer
Manual	Manual for handling of the engine	Test requirer
Specification of the campaign	Manual and campaign specification	Test requirer
Inspection request	Request of interventions on the engine	Test requirer
Measurement request	Specification of the requested measurement points	Test requirer
Test report	Exploitation of engine measurement data	Test requirer
Test report	Exploitation of facility measurement data	Facility operator
Output specification	Definition of diagrams and print-outs	Facility operator
Adjustment instruction	Adjustment list for the facility in test configuration	Facility operator
Test report	Result of interventions and inspections on the engine	Facility operator
Chronology'	Procedure of all test day activities	Facility operator
Work plan	Plan of all activities on the engine	Facility operator

In addition to the test request for the engine, further test objectives can be defined. They concern systems which are not part of the engine but which are activated during a hot run (e.g. the device to gimbal the engine). Such a request is defined separately in a so-called *Passenger Test Request*.

10.2.2 Instruction Manual

The instructions on how to handle the rocket engine are defined in the *instruction manual (specification de mise en ouvre, SMO)*. It covers all activities from the transport to the integration into the facility, the pre- and post-test checks, up to the dismounting and transport back to the client. This document is completed by requests for special inspections and interventions before the upcoming test.

10.2.3 Chronology

The procedure for the overall test execution (*process des operationes des essai)*, short chronology, is the central document of the test day. It lists all work steps and calls for further procedures which are applied in order to prepare the facility and the engine for test and to set both into test configuration. From the start of the chill down phase, another chronology is followed. A chronology for the computer contains the details for the dialog with the MCC system which controls the chill down, the hot run and the set back of the facility and the engine.

10.2.4 Inspection Results

The results of the inspections and the feed back from the interventions requested in the instruction manual are compiled in a test report (*synthèse des résultats obtenus des tests fonctionnels)*. This report is checked and validated by the test requirer before the last dry run, because afterwards no further intervention on the engine is allowed.

10.2.5 Test Report

Concerning the test data exploitation, a division of work was made between the test requirer and the facility operator. The specialists of the test requirer analyse the behaviour of the engine components and compile it into a test report. The behaviour of the facility systems is analysed by the facility operator and documented in another test report. In both reports the behaviour of the systems/components is compared to the objectives in the test request and to the behaviour in previous tests.

10.2.6 Plan of Activities on the Engine

A plan of activities on the engine documents all work procedures which are requested in the instruction manual and applied to the engine. In principle this plan is a list of all applied work instructions during a campaign including date and time of execution. It also contains the name of the person who performed the work and his comments.

Remark 10.1 *The plan of activities for the* ***Vulcain 2*** *engine SN. 206 contains*
83 activities before the first test,
72 of them refer to a predefined written procedure,
35 to 75 activities between the 24 tests of this campaign,
and 58 activities after the last test.

10.2.7 Checklists and Work Procedures

The checklist is the simplest form of a test procedure; it lists the steps taken to perform an activity and gives detailed information. A written test procedure can call on different checklists and other procedures. The set of applicable checklists is divided into one part for work on test facility systems and another for work on the rocket engine itself. The complete set of engine checklists for the *Vulcain 2* on the facility *P5* consists of 281 different checklists (Fig. 10.2).

DLR Deutsches Zentrum für Luft- und Raumfahrt e.V.	**Aufkleben der Temperatursensoren TPACP2, TPACP3 auf den Brennkammerzünder**	Prozedur Nr.: DLR-LA-P5-PR-3103 Version: 1 Datum: 08.09.2009 Seite: 6 von 10

N°	Pos.	Aktion	Bemerkung
		4.2 Kontrolle der Lage des Heizbandes ☐ Kontrolle das das Verbindungsband des TCI-Heizers ca. 15° zur Blindkappe versetzt (siehe Bild 1)	Bild 1: Lage des Verbindungsbandes

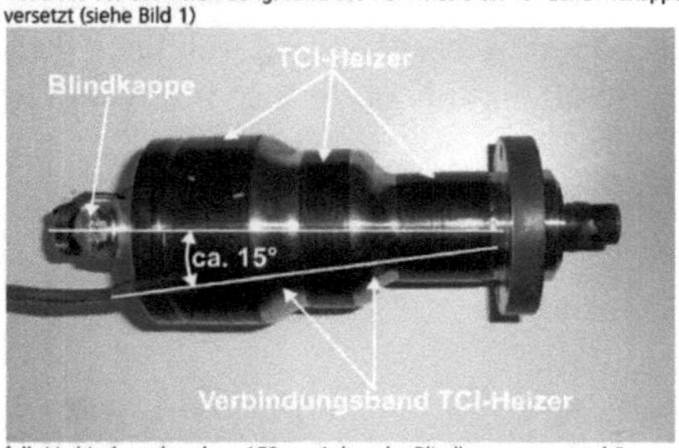

N°	Pos.	Aktion	Bemerkung
		☐ falls Verbindungsband ca. 15° zur Achse der Blindkappe versetzt, können die Temperatursensoren aufgeklebt werden; fortfahren mit Kap. 4.3 und 4.4 ☐ falls die Lage des Verbindungsband zur Achse der Blindkappe abweicht, SNECMA informieren und NCR öffnen; Stoppen der Arbeiten	NCR ....

Fig. 10.2 Snapshot from a work procedure for igniter preparation (Checklist: DLR)

Chapter 11
Test Team

Citation
Testing of rockets is hazardous and is not recommended for hobbyists and amateurs
George P. Sutton [17]

A test facility for a rocket engine is an extraordinary (sometimes even unique) workplace. Most of the systems on the facility are especially designed and their performance is sometimes extreme. The type and amount of fuel and fuel flow can be very specific (e.g. liquid hydrogen). The applied procedures are also specially developed, often having ambitious target (e.g. precision, cleanliness, tightness).

Depending on the size of the test facility a team for test conduction consists of 2–12 persons. For test preparation normally twice as many personnel are necessary and the personnel in stand-by during a test can easily double the size of the team again. The team consists of craftsmen, technicians and engineers from various disciplines. A team member may be the only one in his discipline but nevertheless all tasks on the facility have to be covered by the overall skill of the team. A *training matrix* is a tool to check the capability of the team, listing all the tasks necessary for test conduction and enabling a cross check of the trained persons in the team (Fig. 11.1).

Sufficiently skilled and experienced personal cannot be recruited from other branches in industry. Therefore the formation of a test team takes time and has to be done stepwise. A successive build-up of a team is necessary and this needs the cooperation of the facility management with clients, subcontractors and agencies within the space activities. Not only is the test specimen provided by the client but also a vast set of documentation and rules on how to handle the specimen and how to perform the test. The team has to be trained to perform the test according to this documentation and rules by application of the skill and experience of the team which increases continuously. Any supplier or subcontractor of the facility offers the chance to increase the ability of the test team. Each new system extends the skill of the team and the cooperation with the supplier turns out to be a transfer of *know-how*.

The training at subcontractors or suppliers is usually for a small number of specialists in the team (e.g. handling of vacuum pumps). The internal instructions are

W. Kitsche, *Operation of a Cryogenic Rocket Engine*, Springer Aerospace Technology, DOI 10.1007/978-3-642-10565-4_11, © Springer-Verlag Berlin Heidelberg 2011

Fig. 11.1 Craftsmen during work in a test cell (Photo: DLR)

given mainly to the complete team (Fig. 11.2); often the participation is mandatory (e.g. safety instructions) and is repeated every year. The acquired skills are practiced and maintained in dry runs. Some of the skills are followed up by the authorities (e.g. handling of pyrotechnical elements).

The rules and standards of the European Space Agency are, of course, also applied to the test facilities; they harmonize and facilitate the processes between the teams in space technology. The European Space Agency (ESA) is aware of the need and challenge to maintain the skill and know-how in the teams of their partners and contractors. A main objective of ESA is European guaranteed access to space

Fig. 11.2 Operators in the control room (Image Film: DLR)

(EGAS), the space programs as well as the test plans being designed to match this target.

Last Remark *Instructions, procedures, rules and documents are indispensable on the test facility but the execution is always up to a team which consists of human beings. The most genuine instruction on how to cooperate in a team is given in [18].*

Appendix A
Chamber Valve Characteristic

The position of a valve can be read by the MCC system from end switches; they indicate the full open or close status (Fig. A.1). In a valve switch test characteristic time delays can be identified.

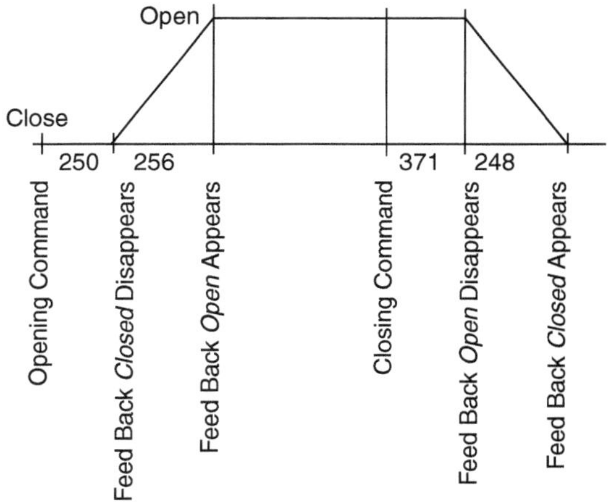

Fig. A.1 Time delay in milliseconds for the manoeuvring of a valve (chamber valve hydrogen of a *Vulcain* engine) (Photo: DLR)

W. Kitsche, *Operation of a Cryogenic Rocket Engine*, Springer Aerospace Technology,
DOI 10.1007/978-3-642-10565-4,

Appendix B
Chamber Igniter Characteristic

The chamber igniter is itself ignited by a pill of powder at its tip. A wire in a short-circuit conducting a current of 6 A for 50 ms ignites the pill. Thus the main powder charge is ignited and the internal pressure (Fig. B.1) of the pyrotechnical element increases. At a pressure of about 120 bar, three membranes which had sealed the element burst. Via three openings at an angle of 120° to each other the hot smoke is injected into the combustion chamber and ignites the fuel/oxidiser mixture. The time taken to reach 80% of the maximum pressure inside the igniter in this context is called the *ignition delay* (Fig. B.1).

Figure B.2 is a series of photos showing the ignition of the fuel/oxidiser mixture and the flame evolution in the combustion chamber. The photos were taken with a high speed camera (2000 frames per second) via a mirror below the engine.

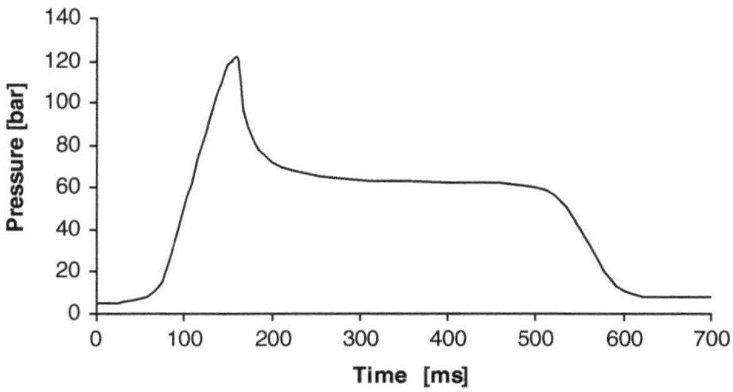

Fig. B.1 Ignition delay of a chamber igniter (Photo: DLR)

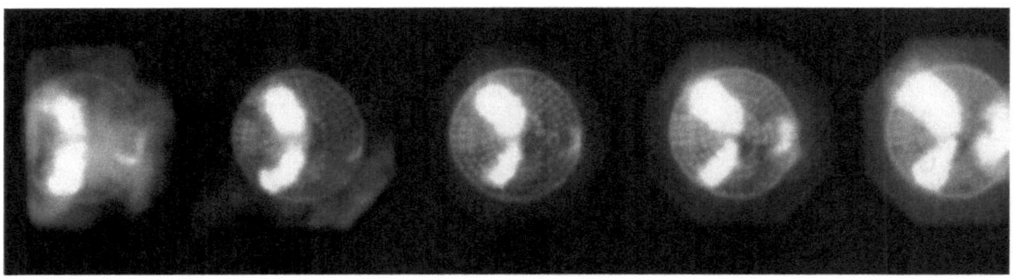

Fig. B.2 Ignition of a *Vulcain*-combustion chamber (Photo: DLR)

Appendix C
Measurement and Correction of the Ovality of a *Vulcain* Nozzle

The measurement and correction of the ovality of the *Vulcain* nozzle has been performed regularly since 1993. The protocol (Fig. C.1) documents such a procedure. The diameter was measured at different circumferential positions (before and after) and with a special clamping device the nozzle exit was reformed into a shape close to the circle.

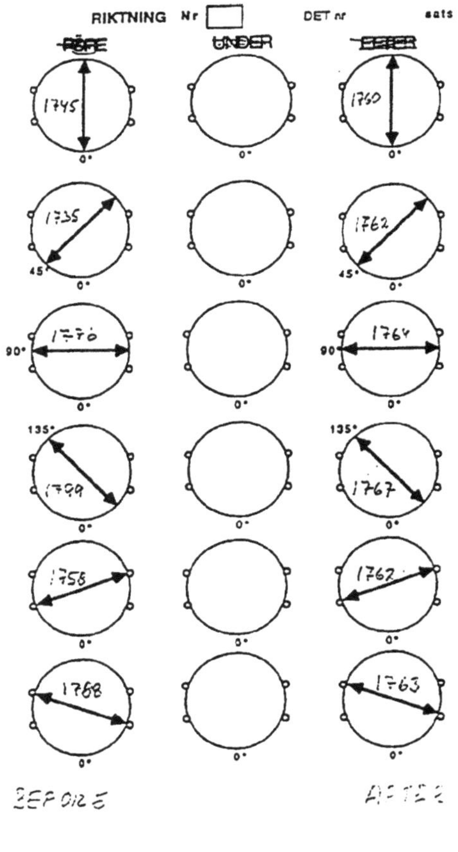

Fig. C.1 Inspection sheet (Inspection sheet: VOLVO)

Appendix D
Compare of Flow Schemes *Vulcain/Vulcain 2*

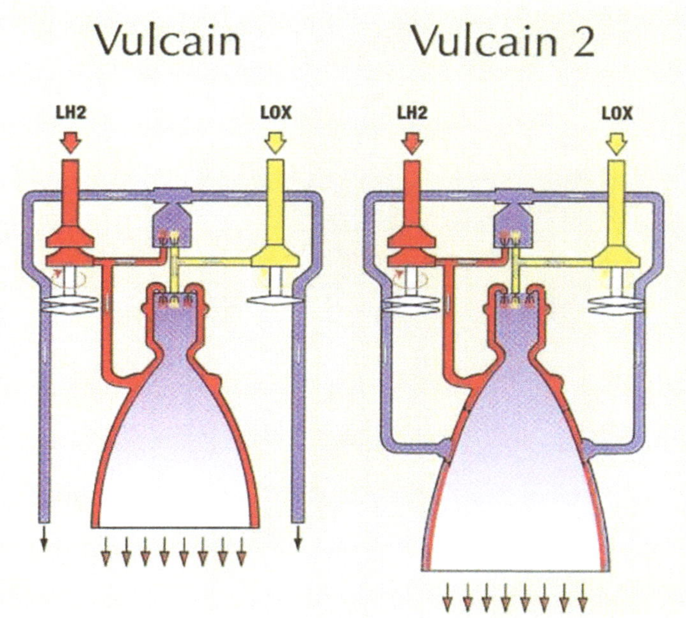

Fig. D.1 Compare of Flow Schemes *Vulcain/Vulcain 2* (Photo: SNECMA)

Appendix E
Flow Scheme of the *Ariane 5* Main Stage

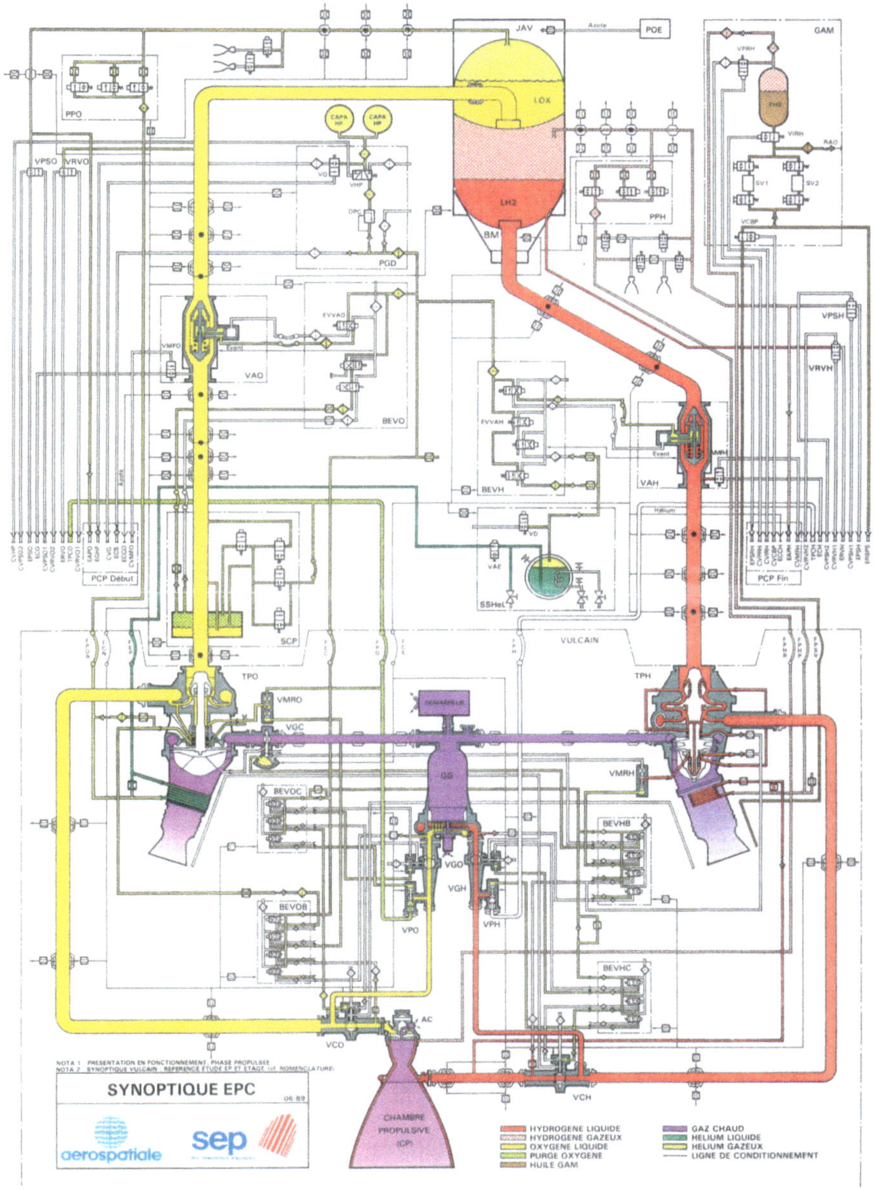

Fig. E.1 Flow scheme of the *Ariane 5* main stage (Photo: SNECMA/EADS (formerly SEP/Aerospatiale))

Appendix F
Flow Scheme of *Vulcain 2*

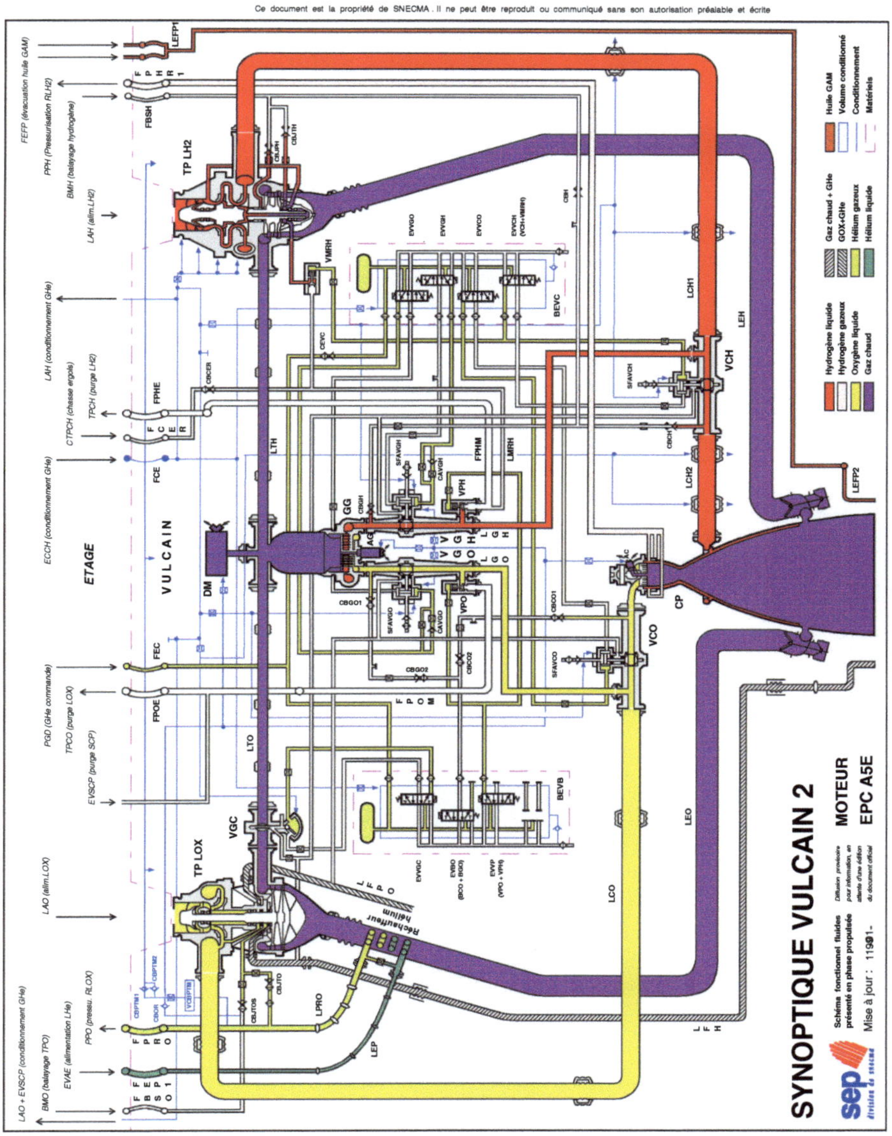

Fig. F.1 Flow scheme of *Vulcain 2* (Photo: SNECMA/EADS (formerly SEP/Aerospatiale))

Appendix G
Flow Scheme of *Vinci*

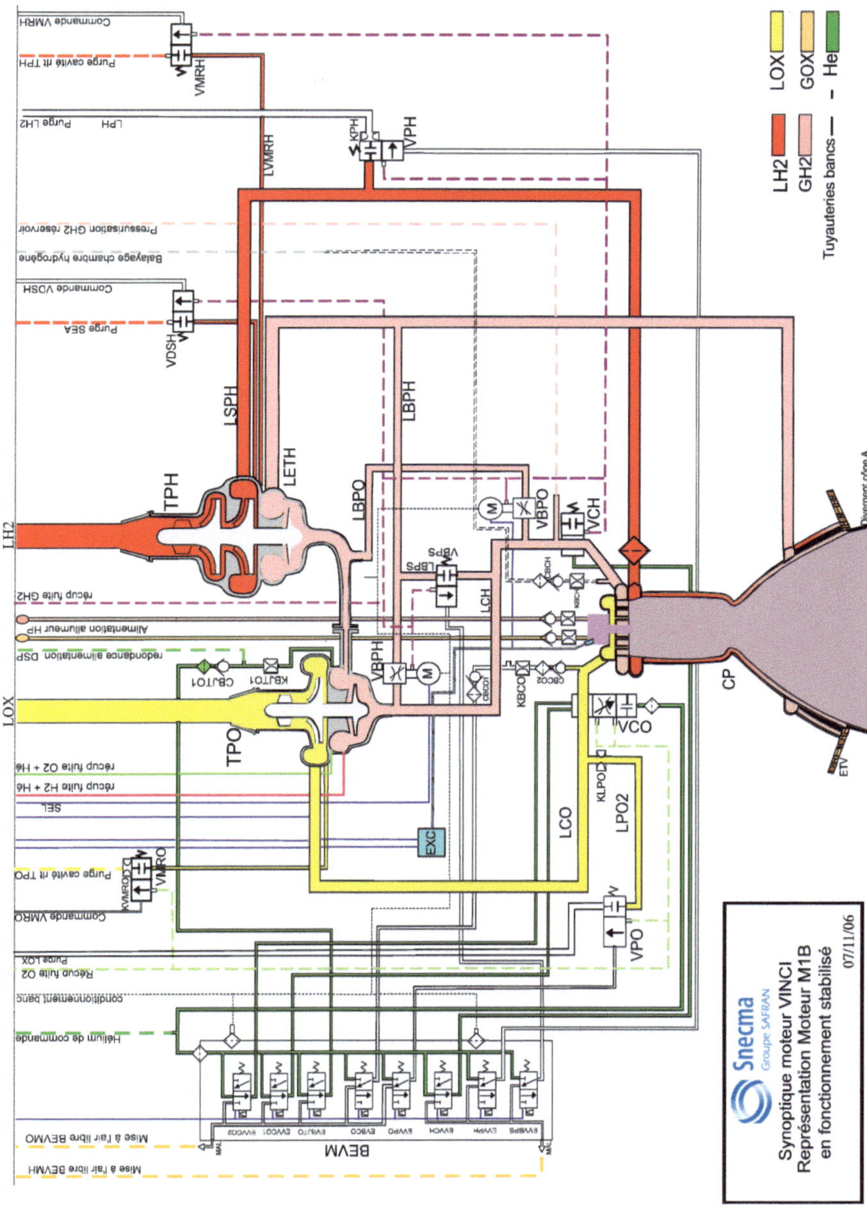

Fig. G.1 Flow scheme of *Vinci* (Photo: SNECMA)

Appendix H
Mixtures of Oxygen and Nitrogen Close to Their Boiling Points

The main component of an oxygen (O_2) supply to a rocket engine is the run tank with liquid O_2. If the tank is pressurised with nitrogen (N_2) the condensation of N_2 at the surface of the liquid O_2 has to be considered. The equilibrium of both fluids close to their boiling point is shown in the phase diagram (Fig. H.1).

For the calculation of the dewpoint and boilingpoint curve (H.1) and (H.2), [16] were used; the vapour pressure $P_i(T)$ was calculated according to [21].

$$\text{Boiling curve } X_1' = \frac{P - P_2(T)}{P_1(T) - P_2(T)} \tag{H.1}$$

$$\text{Dew curve } X_1'' = \frac{-P_1(T)P_2(T)}{P_1(T) - P_2(T)}\frac{1}{P} + \frac{P_1(T)}{P_1(T) - P_2(T)} \tag{H.2}$$

The symbols are:

T temperature of the mixture (here the temperature of the liquid O_2),
P pressure of the mixture (here the pressure in the vessel),
$P_1(T)$ vapour pressure of the first component (here N_2),
$P_2(T)$ vapour pressure of the second component (here O_2),
X_1' liquid mole fraction of N_2,
X_1'' gaseous mole fraction of N_2.

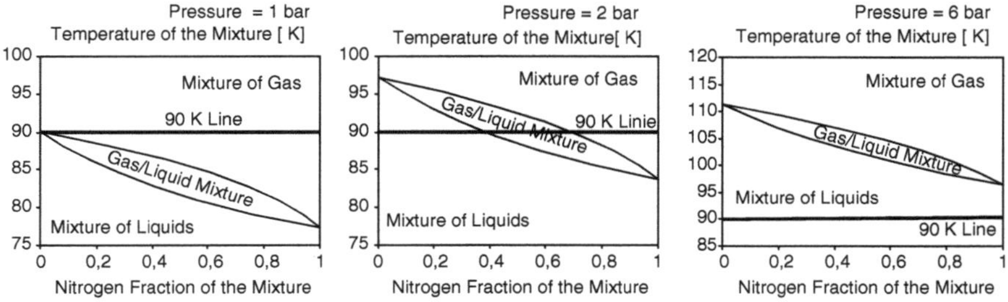

Fig. H.1 Phase diagram of O_2/N_2 mixtures at 1, 2 and 6 bar (Photo: DLR)

In fact the mixture is not homogeneous in time and space and is not at equilibrium. The pressure is defined by the pressurisation system; above the liquid it is constant, below the surface it increases with depth. The temperature around the surface is dominated by the temperature of the liquid oxygen, normally close to 90 K. If the tank pressure is held at 1 bar for a long time only very small fractions of N_2 can be found in the liquid. When the pressurisation is started the condensation at the surface begins and the N_2 fraction increases.

We can assume that the condensed N_2 remains mainly in the upper layers of the liquid (see below). After test, particularly when the pressure is reduced to 1 bar, the N_2 as the more volatile component evaporates again. The typical evaporation rate of a vacuum insulated tank is 0.1% of the inner volume per day. For the oxygen tank (200 m^3) on the facility *P5* that means 200 L per day. Hence it takes 7 days to re-evaporate the N_2. But in the normal test cycle the tank is refilled the day after the test. Due to this dilution the concentration of N_2 decreases to a tenth part, which is a far more drastic decrease than that due to the evaporation. Nevertheless, for the engine test the N_2 fraction in the oxygen means pollution. A sample taken during a normal test period had a fraction of 0.36% of N_2 in O_2.

Diffusion of Liquid N_2 in Liquid O_2

In order to check whether a fast diffusion can distribute the N_2 fraction all over the tank, we have to look for the solution of the *diffusion equation* (H.3) at the applicable conditions. We also need the coefficient of N_2 diffusion in liquid O_2. The diffusion equation has the same mathematical structure as the *heat conduction equation* (H.5) for which we find the general solution (H.6) in [16] and transfer it to the diffusion equation. For the *diffusion coefficient D* no measured value could be found in the literature or on the internet; therefore the Stokes-*Einstein*-equation (H.4) was applied.

$$\text{Diffusion equation } \frac{\partial c}{\partial t} = D \frac{\partial^2 c}{\partial x_i \partial x_j} \tag{H.3}$$

where c is the concentration, t the time, and x_i the position vector (or, more correctly, the coefficients of the components of the position vector).

The *Stokes-Einstein* equation gives us

$$D = \frac{K_B T}{6\pi\, \mu\, R_0} = 97.6 \times 10^{-9}\,\text{Ns/m}^2 \tag{H.4}$$

where $K_B = 1.38 \times 10^{-23}$ as the *Boltzmann* number, $T = 90\,\text{K}$ as temperature, $\mu = 6.5 \times 10^{-6}\,\text{Ns/m}^2$ as the dynamic viscosity (of O_2 at 90 K, 6 bar) and $R_0 = 10.4 \times 10^{-11}$ m as particle radius (of N_2).

We obtain the general solution for the concentration $c(x, t)$ in the one dimensional case:

$$c(x,t) = c_0 + c_1 x + \sum_n \left\{\exp\left(-D\,a_n^2 t\right)(A_n\,\cos(a_n\,x) + B_n\,\sin(a_n\,x))\right\} \quad \text{(H.5)}$$

with the constants c_0, c_1, a_n, A_n, B_n.

The importance of N_2 diffusion in the O_2 tank is checked in an example. Assume at time $t = 0$ a pure layer of liquid N_2 is situated above a pure layer of liquid O_2. Both layers have the same thickness, together $d = 100\,\text{mm}$. (this complies with the N_2 mass which normally condensates in our tank):

$$c(x,t) = \frac{c_{\max}}{2} - 2\,c_{\max} \sum_{\substack{n>0 \\ \text{odd}}} \left\{\exp\left(-D\left(\frac{n\pi}{d}\right)^2 t\right) \frac{1}{n\pi}\sin\left(\frac{n\pi}{d}x\right)\right\} \quad \text{(H.6)}$$

The statements at (H.5) and (H.6) are transferred from the solution of a heat conduction problem, here the adiabatic bar [16].

We can see (Fig. H.2) that the relative concentration $c/c_{\max}$ after 2 h of diffusion is still far from the homogeneous (50% of $c_{\max}$ at each point) state. Hence we can conclude that the process of diffusion cannot really distribute the N_2 in the tank within the relevant small period of time. Pressurisation, depressurisation and filling provide a much stronger mixing process of the two components.

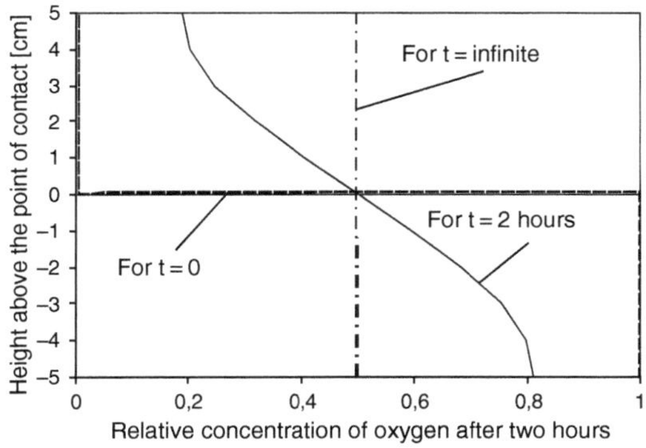

Fig. H.2 Diffusion in a layer of N_2 and O_2 (Photo: DLR)

Appendix I
Jet Pump

A fluid flow engine which increases the total pressure of a fluid is called a pump (for a liquid) or a compressor (for a gas) (Fig. I.1). A very simple engine of this category is the jet pump. It has a nozzle to create a jet (gas or liquid) of high velocity but low pressure. Due to the low pressure the jet is able to suck in fluids from a somewhat higher pressure level. This fluid is carried along and energy is transferred from the jet to the fluid. Behind a mixing passage the flow is guided in such a manner that the kinetic energy is transformed into pressure energy, this pressure being higher than the original pressure of the fluid. Hence we rightly state that the fluid carried along is compressed or pumped. The jet pump has no moving or rotating parts, it can provide high power despite small dimensions and it is very suitable to suck off fluids from low pressure regions. For short term operation (e.g. 1 h) on a test facility the jet pump is preferred in comparison to other pumps (e.g. rotary vane pump) due to lower investment and maintenance costs.

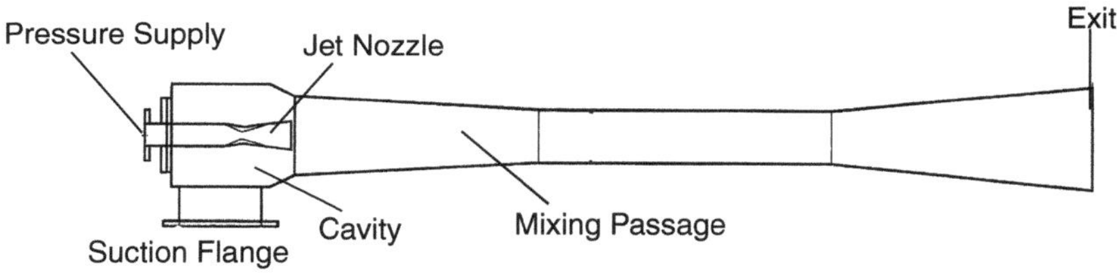

Fig. I.1 Gas jet pump on the facility *P5* (Photo: DLR)

The convergent/divergent jet nozzle is driven at high pressure (e.g. 40 bar) and provides a supersonic jet of low static pressure. The jet increases in diameter and impinges the inner wall of the mixing passage. Up to that point it sucks in gas from the cavity and creates here a pressure decrease. At the impingement point (respectively circle) a sonic shock or a series of shock patterns occurs. Via the suction flange further gas is sucked in. Behind the mixing passage the jet is further compressed and leaves the jet pump as a subsonic jet.

For the design of a jet pump and for the computation of the operational parameters the equations of balance for energy and mass are applied, as well as functions of gas dynamics and empirical coefficients.

One operational mode of a jet pump is the *zero-suction-mode*. This mode means an operation at closed suction valves; the cavity is evacuated and no suction flow is possible. We study a full supplied jet pump (reference point) in this mode. As normal the jet impinges the inner wall, the *Mach* number M_A in this point depending on the diameter of the jet at that point. The sonic shock causes a strong loss of total pressure. The ratio of total pressure behind the shock to the ambient pressure is equivalent to the pressure ratio (static/total) at the exit. For computation purpose we make a variation of M_A, compute the pressure loss across the shock and compute the pressure ratio at the exit (Fig. I.2). In the solution M_A has to match the correct static pressure at the exit which adapts to the ambient pressure. The area and the static pressure of the jet at the impingement point can be computed as well from M_A and hence follows the suction pressure in zero-suction-mode.

For the supply of the jet pump with 14 kg/s N_2 at 288 K the computation of the zero-suction-mode is summarized in Table I.1. Even without knowledge of the loss coefficients in the supersonic jet and in the mixing passage, we have good agreement of the computation with the measurement during operation.

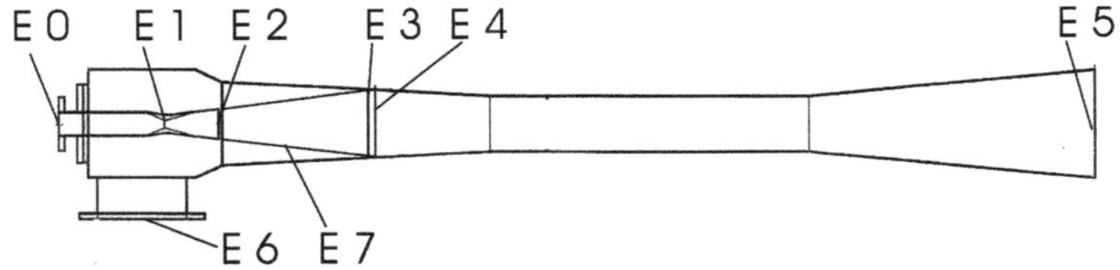

Fig. I.2 Computation sections of the jet pump (Photo: DLR)

In the normal operation the suction flow has to be considered. The lower the suction pressure of the pump the higher the performance. Flow and pressure can be shown as a characteristic of the jet pump. On the other hand the flow also depends on the leak ratio of the connected device (e.g. a vacuum chamber) and on the pressure loss in the line between. The behaviour of this device is also given as a characteristic. The intersection of both characteristics is the reference point of the jet pump (Fig. I.3).

To calculate the properties of the mixture of jet and suction flow (Table I.2) we apply the balance of mass and energy and assume that the entropy of the mixture is

Table I.1 Gas dynamic parameters in zero-suction-mode

Section		Static pressure [bar]	Total pressure [bar]	Area [m^2]	Mach number
E 0	Supply	40	40		
E 1	Nozzle throat	21.13	40	0.0015	1
E 2	Nozzle exit	0.1924	40	0.0201	4.24
E 3	Directly before the shock	0.02	40	0.0935	6.232
E 4	Directly before the shock	0.9052	1.012	0.0935	0.4023
E 5	Directly behind the shock	1	1.012	0.283	0.13
E 6	Suction flange	0.02			

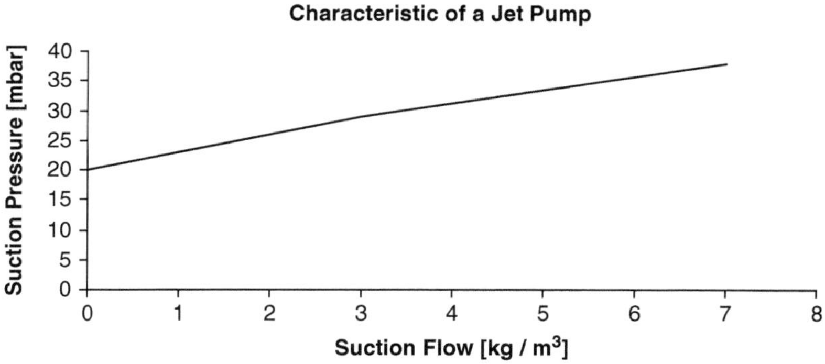

Fig. I.3 Characteristic of a jet pump (Photo: DLR)

Table I.2 Gas dynamic parameters at 3 kg/s suction flow

Section		Static pressure [bar]	Total pressure [bar]	Area [m^2]	Mach number	Mass flow [kg/s]
E 0	Supply	40	40			14
E 1	Nozzle throat	21.13	40	0.0015	1	14
E 2	Nozzle exit	0.1924	40	0.0201	4.24	14
E 3	Directly before the shock	0.0287	20.9	0.1083	5.28	17
E 4	Directly before the shock	0.93	1.04	0.1083	0.403	17
E 5	Directly behind the shock	1	1.04	0.283	0.2375	17
E 6	Suction flange		1.04			3
E 7	Surface of the jet		1.04			

equal to the sum of entropy of both flows. Indeed, we know the impulse of the mass flow but not the forces on the inner surface and therefore we do not use the balance of impulse.

$$\left(\dot{m} c_p T_G\right)_{\text{Jet}} + \left(\dot{m} c_p T_G\right)_{\text{Suction Flow}} = \left(\dot{m} c_p T_G\right)_{\text{Mixture}} \tag{I.1}$$

$$\dot{m}_{\text{Jet}} + \dot{m}_{\text{Suction Flow}} = \dot{m}_{\text{Mixture}} \tag{I.2}$$

$$(\dot{m}\, s)_{\text{Jet}} + (\dot{m}\, s)_{\text{Suction Flow}} = (\dot{m}\, s)_{\text{Mixture}} \tag{I.3}$$

$$s = c_P \ \ln\left(T_{\text{total}}/T_{\text{reference}}\right) - R_{N_2} \ \ln\left(P_{\text{total}}/P_{\text{reference}}\right) \tag{I.4}$$

The specific heat c_P of both components (N_2 and air) is considered as equal, $\dot{m}$ is the mass flow and T_{total} and P_{total} are total temperature and pressure of the gas. Total in contrast to static is used her in the sense of gas dynamics (see Remark I.1). The specific entropy s refers to $T_{\text{reference}} = 288$ K and $P_{\text{reference}} = 1$ bar. R_{N_2} is the specific gas constant of nitrogen.

Remark I.1 *The technicians on the test facility also use the term* ***static*** *for a pneumatic system when no consumer load is active (no consumption). The system is, e.g. adjusted to 20 bar before the consumer is activated, the pressure may drop then, e.g. by 2 bar when the consumers are active. This use of the term must not be confused with the context of gas dynamics. In general all pressure values in the* ***Ariane*** *program (another context of terms) are given as absolute values relative to the vacuum pressure of 0 bar.*

Because jet and suction flow have both $T_{\text{total}} = 288\,K$ *and* $P_{\text{total suction flow}} = 1$ *bar we can conclude by (I.3) and (I.4):*

$$\left(\dot{m}\,\ln\left(\frac{P_{\text{total}}}{P_{\text{reference}}}\right)\right)_{\text{jet}} = \left(\dot{m}\,\ln\left(\frac{P_{\text{total}}}{P_{\text{reference}}}\right)\right)_{\text{mixture}} \tag{I.5}$$

From (I.5) we can compute the total pressure before the shock and with the total temperature and mass flow we continue our computation analogue to the zero-suction-case.

The calculated suction pressure was 28.7 mbar, and 115 mbar was measured. The difference is only 86 mbar (8.6% relative to 1 bar ambient pressure) but, on the other hand, the result of this nominal case is just in the order of the measured values. However, the system was designed for a pressure of 200 mbar in the casing, and the measurement was taken at 300 mbar.

For isentropic compression of 3 kg/s from 29 mbar to 1 bar a power of 1512 kW is required. The supply of the jet pump (40 bar, 14 kg/s) is 7536 kW which means an efficiency of $\eta = 1512/7536 = 0,2$. This efficiency is much less than for any other hydraulic/mechanic pump, the reason for the low efficiency being the high entropy production across the sonic shock.

The characteristics of the jet pump were measured in pre-tests in which the end of the suction line was closed except for a combination of orifices. The casing was not connected to the jet pump.

The mass flow during operation depends on the leakage of the casing. As soon as the casing pressure is below 0.52 bar a critical pressure ratio is reached at the points of leakage and the leak flow does not increase further more (except if the leak area increases). The desired pressure in the casing can be adjusted by the regulation valves which quasi-create a desired pressure loss in the line. Hence we have

$$P_{\text{casing}} = P_{\text{suction}} + \Delta P_{\text{line}} + \Delta P_{\text{valve}} \tag{I.6}$$

The pressure loss at a sonic/subsonic transfer is not avoidable. The highest loss can be assumed across a normal shock (perpendicular to the flow), planar oblique shocks (wedge like) having less pressure loss and the oblique shock at the tip of a cone again having less pressure loss. The loss across shocks at the inner wall of a tube (as we have it in a jet pump) is a bit less than at a normal shock but still significantly higher than in a wedge like flow. In particular in a jet pump of high power (e.g. for the vacuum chamber of an upper stage engine) it is desirable to reduce the pressure losses across the sonic shocks. For that purpose an rotationally symmetric centre body is integrated behind the mixing passage. The sharp cone has the perfect gas dynamic performance; it causes a slight deviation of the flow and decelerates the flow to a low *Mach* number. But the tip would have to bear a very high thermal load and therefore the centre body normally has a blunt nose which creates a detached shock in front of it.

Appendix J
Fluids of the Test Process

For the operation of a test facility it is of essential importance to consider carefully the properties of the fluids used within the operation (test). We use the term fluids because some substances are used in liquid as well as in gaseous form. Further on the possibility of freezing of the fluids must also be checked. The most important fluids on a test facility for a cryogenic rocket engine are listed in Table J.1.

At ambient condition (15 °C, 1 bar) the hydrogen and helium is much lighter than air the gas climbs up and may accumulate at the ceiling of a closed room.

Remark J.1 *In the design of the test cell of the facility* ***P5*** *the roof is lower on the side of the tower of the building and higher the on the outward side. Due to this design the ascending hydrogen is deflected from the building. On the other hand, unfortunately, rain water on the roof is directed towards the building. Any holes in the test cell roof directly cause problems because the penetrating water is collected in the thrust cone like in a giant funnel and than directed to the engine where it can cause failures on electrical and electronic components.*

Even in the liquid state (20 K, 1 bar) hydrogen is relative light, its specific density equals the density of rock wool. The specific density of liquid oxygen is of the order of water.

Remark J.2 *On the facility* ***P5*** *in November 1991 severe damage occurred due to a broken water line. The chute for the hydrogen tank filled up temporarily almost to the top. The tank was at that moment not far from floating. The empty steel tank of 200 tons and its hydrogen filling of approximately 30 tons were much less than its displacement of more than 600 tons of water. Thanks to the fact that the tank was not totally covered by water and that the connections to the facility did not break, no greater damage to the test facility occurred.*

Remark J.3 *The routing of the oxygen feed line is vertical for several metres and is equipped with flow turbines. To fill the line from the top is harmful for the turbines, and therefore an alternative filling procedure was developed. In order to avoid a* ***water hammer*** *in the oxygen feed the closing times for the valves in this line have to be defined with extra care.*

Table J.1 Fluids in the test process of a cryogenic test facility

		Hydrogen	Helium	Oxygen	Nitrogen
Specific density at 15 °C and 1 bar	kg/m^3	0.0841	0.167	1.337	1.17
Ratio of the specific heat Cp/Cv = κ		1.41	1.66	1.4	1.4
Specific density at the boiling point	kg/m^3	70	125	1140	810
Molecular weight M	kg/kmol	2.016	4.003	32	28.01
Specific gas constant R_S	J/(kg K)	4124.16	2077.02	259.82	296.83
Triple point	bar	0.072	–	0.0015	0.125
	kg/m^3	80 or 0.13	–	1306 or 0.01	867 or 0.68
	K	13.95	–	54.4	63.15
Critical point	bar	13.16	2.3	50.9	33.98
	kg/m^3	31.57	70	405.8	281
	K	33.2	5.2	154.8	126.3
Evaporation heat r at 1 bar	kJ/kg	460	20.59	238.7	198.2
Melting point at 1 bar T_m	K	14	1	54.8	63.3
Boiling point at 1 bar T_b	K	20.4	4.2	90.2	77.3
Specific heat at constant pressure c_p	kJ/(kg K)	14.32	5.23	0.917	1.038

		Air	Propane	Water	hydraulic oil
Specific density at 15 °C and 1 bar	kg/m^3	1.21	1.88	1000	869
Ratio of the specific heat Cp/Cv = κ		1.4	1.14	1.33	
Specific density at the boiling point	kg/m^3		581	1000	
Molecular weight M	kg/kmol		44.1	18	
Specific gas constant R_S	J/(kg K)	287.00	188.53	461.91	
Triple point	bar			0.0061	
	kg/m^3			1000 or 0.005	
	K			273	
Critical point	bar		42.42	221	
	kg/m^3				
	K		370	647	
Evaporation heat r at 1 bar	kJ/kg		426.5	2258	
Melting point at 1 bar	K		86.5	273.15	(213)
Boiling point at 1 bar	K		231	373.15	
Specific heat at constant pressure	kJ/(kg K)	1.005	1.595	1860	

When heat is introduced into a cryogenic fluid, and this is mostly the case because there is no perfect insulation, evaporation and a pressure increase occurs. Therefore any segment for a cryogenic fluid has to be equipped with safety components (burst disc, safety valve) and the process has to be conducted in such a manner that the fluid is never locked.

In cases of direct contact of different fluids the reactivity (e.g. detonating gas) has to be considered. In cases of extreme temperature differences condensation and icing has to be expected. These effects can also occur on the outer surface of tubes and engine components.

Remark J.4 *During the first start up of a sub-system involving liquid helium, at some tubes a lot of air was liquefied. Besides water, liquid* N_2 *in fact also rained down from the tubes.*

Remark J.5 *In a combustion chamber for research purpose running on liquid hydrogen/oxygen, an annular icing around the injection elements was detected. Fuel and oxidiser are ducted through the injection plate and thus keep it at a very low temperature despite the combustion downstream. In the recirculation zone directly below the injection plate* ***(face plate)*** *there is of course water as the product of the combustion. This water, cooled down, deposits as ice at the face plate.*

If a gas gets in contact with a cryogenic fluid, condensation and mixing can occur (see Appendix H). Due to the contact a pressurisation gas normally cools down, increases in density and decreases in pressure. Sometimes the term *collapse factor* is used to describe this effect but this is no mysterious phenomenon and it is not necessary to measure the factor empirically; it can be explained and calculated within thermodynamics.

Another term used for an effect in the context of cryogenic fluids is *cryo-pumping*. Remark 5.4 describes impressively what happens when a gas comes into contact with a surface whose temperature is lower than the saturation temperature of the fluid. The gas increases in density, condenses to liquid and the pressure decreases if this occurs in a closed cavity, or further gas is sucked in if the cavity is open.

The most meaningful diagram concerning the thermodynamic state of a fluid is the *enthalpy-entropy-diagram* (h-s-diagram, Fig. J.1). In particular, a status or process close to the two-phase area (liquid/gaseous) can be visualised properly in the h-s diagram. At normal condition (1 bar, 0 °C) hydrogen, oxygen and nitrogen are far from their two-phase area. But the processes in a cryogenic rocket engine are running close to that area or traverse the two-phase area.

Extremely high temperatures (e.g. the H_2 fraction in the exhaust of the rocket) are not included in Fig. J.1. The properties of hydrogen in this state are given in Table J.2.

At extremely high pressures (e.g. in a bottle at 250 bar, 0 °C) the gas has no *ideal* behaviour any more. In the given example the compressibility factor $z = pv/(R_S T)$ is 1.13.

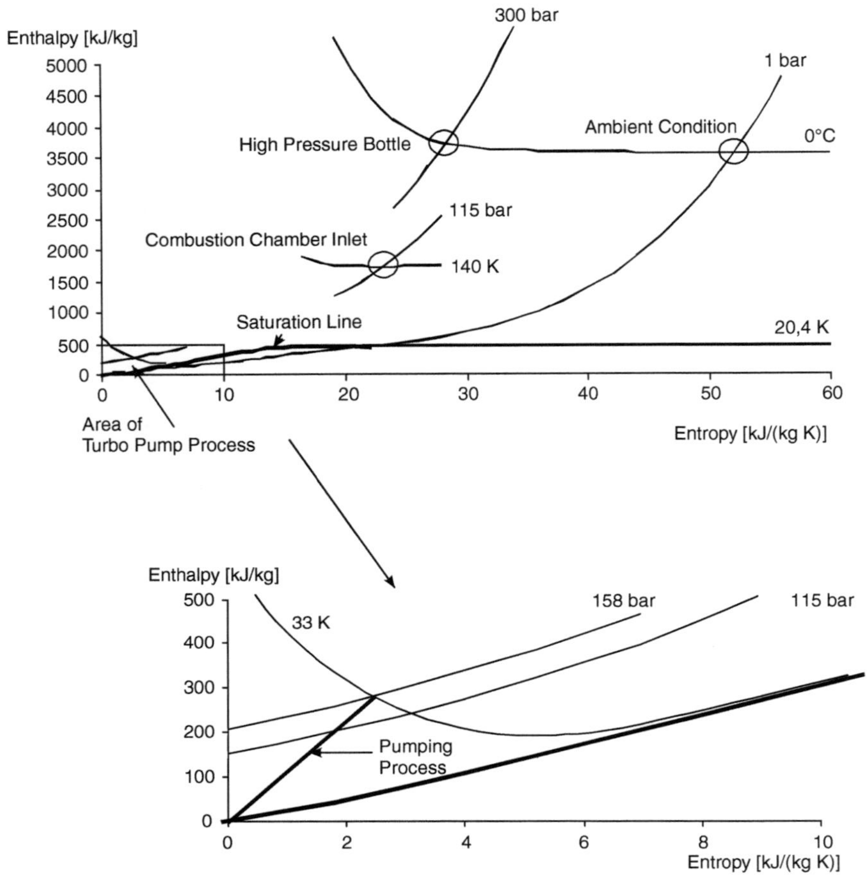

Fig. J.1 Enthalpy-entropy diagram of hydrogen (Photo: DLR)

Table J.2 Hydrogen at normal conditions and at high temperature [5]

Pressure	bar	1	100
Temperature	K	273.15	3000
Density	kg/m^3	0.0887	08117×10^{-3}
spec. Enthalpy	kJ/kg	3573	48460
spec. Entropy	kJ/(kg k)	52	81.67
C_p	kJ/(kg K)	14	18.49
C_v	kJ/(kg K)	10	14.35
Speed of sound	m / s	1261	4026
$\kappa = C_p/C_v$		1.41	1.29

Appendix K
Pressure Transducer

In the typical pressure transducer on the test facility and on the rocket engine, *strain gauges* are widely used to transform pressure (as the physical parameter) into voltage (as an electrical parameter). The strain gauge is a flexible membrane with thin wires glued to the surface. The strain gauges are arranged in the manner of a *Wheatstone bridge*. The membrane is deformed according to the pressure and the deformation changes the resistance of the strain gauges (Figs. K.2–K.4) and hence a measurement voltage is available on the *Wheatstone* bridge. As an example, in Fig. K.1 the strain gauges (resistors) R1, R4 are expanded and R2, R3 are compressed if the membrane is deformed.

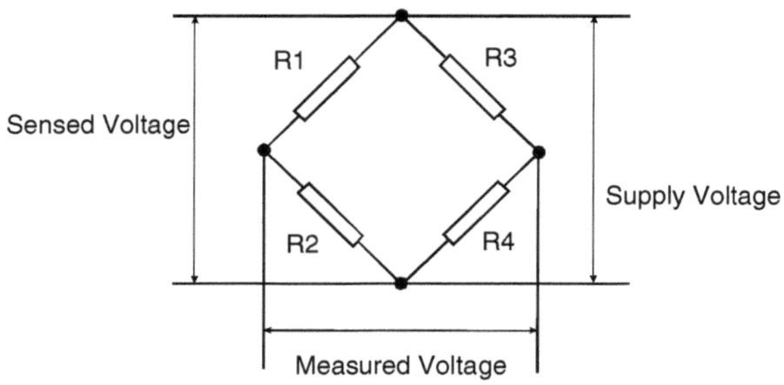

Fig. K.1 Strain gauge in a *Wheatstone* bridge measurement element (complementary resistances for temperature compensation are not shown) (Photo: DLR)

The measured voltage is proportional to the supply voltage. On test facilities the supply is normally far from the sensor and a loss in the cable must be considered. Therefore the effective supply voltage is also measured on the sensor by means of a so-called *sense line*. That means there are six wires (plus screen) in one sensor cable (Table K.1). A higher resistance of the sensor would reduce or avoid a voltage loss but it has the disadvantage that dynamic pressures cannot be measured because the capacity of the cable combined with a high resistance of the sensor causes a considerable damping in the measurement chain.

Fig. K.2 Strain gauge in a pressure transducer (Photo: DLR)

Fig. K.3 Disassembled pressure transducer (Photo: DLR)

Fig. K.4 Sensor with cut casing (Photo: DLR)

Table K.1 Typical parameters of a pressure transducer

Supply voltage	12	V
Cable voltage loss	2	V
Sensed voltage	10	V
Resistance of the cable	60	Ohm
Resistance of the sensor	300	Ohm
Measured voltage at full signal	20	mV

Appendix L
Measurement Chain

The design of a measurement chain depends on the parameter to be measured (temperature, pressure, vibration etc.) and on the measurement mode (range, acquisition rate, precision etc.).

Normally the chain from the sensor into the computer has several plugged and fixed cable connections and has at least one unit for amplification and signal adjustment. Normally the analogue signal is converted at the entrance of the computer. Inside the computer the signal is treated again for different purposes (archiving, regulation, display, monitoring).

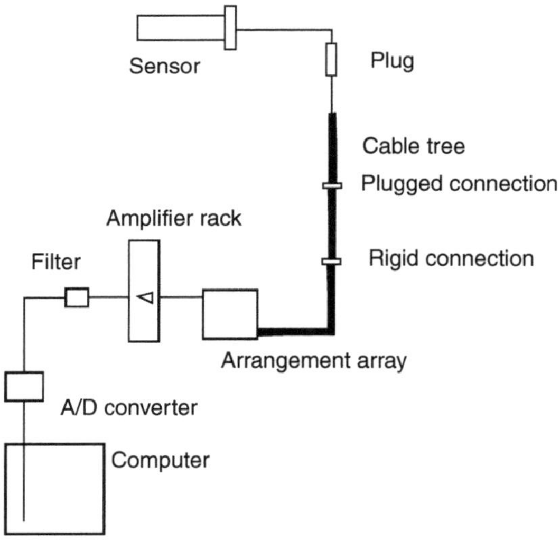

Fig. L.1 Typical measurement chain between a pressure transducer and the computer system (Photo: DLR)

Appendix M
Valve Control Circuit

Most the valves on the facility as well as on the rocket engine are opened/closed by means of a pneumatic actuator. The activation of the actuator is again controlled by a pilot valve, an electrically driven open/close valve which switches the in/outlet of the actuator to a pressure gas source or to a venting line (atmosphere). The electrical actuation of the pilot valve is realised by a chain of electrical components between the valve and the control computer.

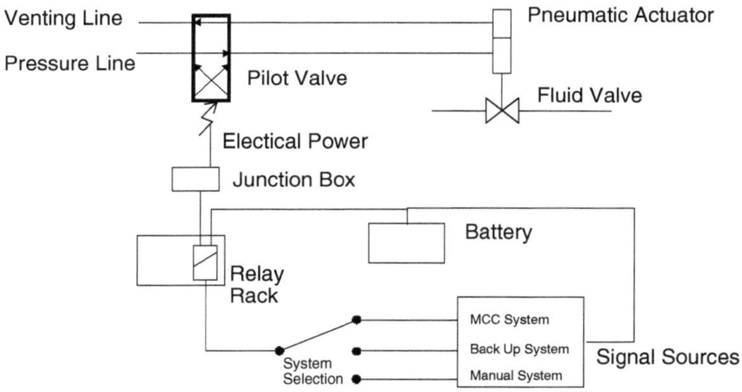

Fig. M.1 Typical valve control circuit for an open/close valve (Photo: DLR)

The outer tube of the vacuum sections (Fig. M.3) is a rigid, welded or well sealed tube of stainless steal. The outside of the conventional insulation (Fig. M.4) looks almost the same, but here we have a thin aluminium cover which protects the hard foam insulation.

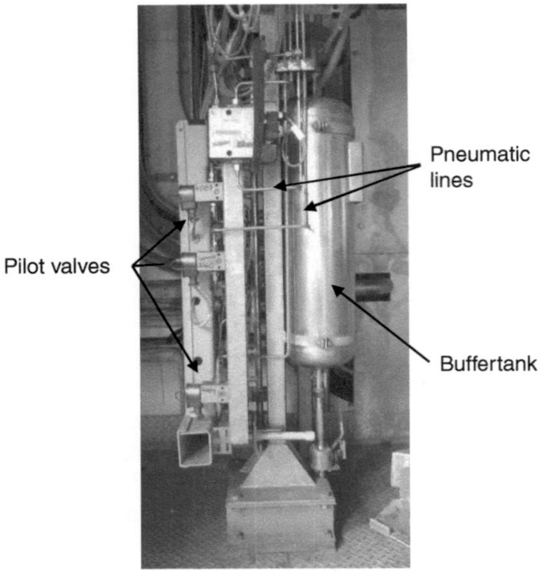

Fig. M.2 Rack of pilot valves (Photo: DLR)

Fig. M.3 Automatic valves for cryogenic lines (integrated in a vacuum box) (Photo: DLR)

Fig. M.4 Automatic valves for venting lines (integrated in "conventional" insulation) (Photo: DLR)

Appendix N
Oxygen Detector

Design

The measuring cell consists of a plastic casing which houses two electrodes emerged into an electrolyte. The cathode is a gold plated grid, the anode is a cylinder made of sintered lead. The tightness of the component is good but gas tightness is guaranteed by a thin Teflon diaphragm.

Fig. N.1 Gas analyser rack with oxygen detectors (blue casing) (Photo: DLR)

Measurement Principle

When the anode comes into contact with oxygen a reduction – oxidation reaction is initiated. Due to the reaction a potential difference (voltage) is created between the electrodes. The voltage is proportional to the partial pressure of the oxygen. Because the cell is influenced by temperature an adjustment by means of a thermistor is necessary. The voltage at the contacts is amplified and displayed.

References

1. S.A Durteste, Transient model of the VINCI cryogenic upper stage rocket engine, *AIAA Joint Propulsion Conference*, Cincinnati, 2007
2. J. Gastal, J.R.L. Barton, VULCAIN: A cryogenic engine for ARIANE 5, lecture series 1993-01 von Karman Institute for Fluid Dynamics
3. A. Haberzettl et al., VULCAIN 2 flight load simulation device, *EUCASS European Conference for Aerospace Sciences*, Moscow, 2005
4. D.T. Harrje, F.H. Reardon, Liquid propellant rocket combustion instability, *NASA SP-194 National Aerospace and Space Administration*, U.S.A, 1972
5. R.C. Hendricks et al., NASA TN D-7808 Cleveland, OH, 1975
6. C. Hujeux, W. Kitsche, Evolution of the rocket engine testing process, *AAAF Association Aeronautique et Astronautique de France*, Versailles, 2002
7. INERIS Retour d'expérience issu de la mise en ouvre d'un réservoir d'hydrogène liquide haut pression, Journée technique, 07.10.2003
8. Exploitation of various internet websites
9. P. James,Technological readiness of the vinci expander engine, *IAC International Astronautical Congress*, Glasgow, 2008
10. W.H. Kitsche, Simulation of flight conditions on a test facility for rocket engines, *EUCASS European Conference for Aerospace Sciences*, Brussels, 2007
11. W.H. Kitsche, Pollution control on a test facility for a cryogenic rocket engine, *EUCASS European Conference for Aerospace Sciences*, Versailles, 2009
12. K. Koch, Analysis of signals from an unique ground-truth infrasound source observed at IMS station IS26 in southern germany, *Pure Applied Geophysics*, **167**, 401–412, Basel, 2010
13. C.R. Koppel et al., A platform satellite modelling with ecosimpro: simulation results, *AIAA Joint Propulsion Conference*, Denver, 2009, AIAA 2009, 5418
14. P. Magnant, B. Juery, and N. Chazal, PF52 test facility for cryogenic engines and subsystems, *SpaceOps Conference*, Huntsville Alabama 2010, AIAA 2010-2253
15. R.E. Martin, Atlas II and IIA analyses and environments validation, *Acta Astronautica* **35**(12), 1995
16. I. Müller, *Grundzüge der Thermodynamik mit historischen Anmerkungen*, 3. Auflage (Springer, Berlin, 2001)
17. G. Ordonneau, F. Lévy, Low frequency oscillation phenomena during VULCAIN shutdown transient, *AIAA Joint Propulsion Conference*, Salt Lake City, 2001
18. Holy Spirit, *Bible* (Christian Church, Worldwide)
19. G.P. Sutton, *Rocket Propulsion Elements* (Wiley, New York, NY, 1986)
20. M. Williamson, *Dictionary of Space Technology* (Adam Hilger, New York, NY, 1990)
21. W. Wagner, *Multifluid Package* (Ruhruniversität Bochum, 2003)

Index

About the Author

Wolfgang Kitsche works as a senior test leader at the test centre for rocket engines of the German Aerospace Center (DLR), Institute of Space Propulsion, Lampoldshausen, Germany. Impressed by the manned space flights of NASA in the 1960s he decided as a young boy to become an aircraft engineer. He studied aerospace engineering at the Technical University of Berlin and focussed on propulsion and thermodynamics. Several internships in the field of physics of propulsion at VFW-Fokker, Bremen and some years work on turbo machinery at Borsig, Berlin extended his theoretical knowledge before he entered his current position, which he once called in an interview *a present from heaven* (see Job 1.21).

MIX
Papier aus verantwortungsvollen Quellen
Paper from responsible sources
FSC® C105338

If you have any concerns about our products,
you can contact us on
ProductSafety@springernature.com

In case Publisher is established outside the EU,
the EU authorized representative is:
Springer Nature Customer Service Center GmbH
Europaplatz 3, 69115 Heidelberg, Germany

Printed by Libri Plureos GmbH
in Hamburg, Germany